T0186530

The Bioorganic Chemistry of Enzymatic Catalysis: An Homage to Myron L. Bender

Editors

Valerian T. D'Souza, Ph.D.
Department of Chemistry
University of Missouri-St. Louis
St. Louis, Missouri

and

Joseph Feder, Ph.D.
Invitron Corporation
St. Louis, Missouri

CRC Press
Boca Raton Ann Arbor London

Library of Congress Cataloging-in-Publication Data

The Bioorganic chemistry of enzymatic catalysis: an homage to Myron
 L. Bender / editors, Valerian T. D'Souza and Joseph Feder.
      p.   cm.
   Includes bibliographical references and index.
   ISBN 0-8493-6823-5
   1. Enzymes. 2. Bioorganic chemistry. 3. Bender, Myron L., 1924-1988.
      I. Bender, Myron L., 1924-1988. II. D'Souza, Valerian T.
   III. Feder, Joseph.
QP601.B485   1991
574.19'25--dc20   91-30893
                                                                  CIP

   Direct all inquiries to CRC Press, Inc., 2000 Corporate Blvd., N.W., Boca Raton, Florida
33431.

© 1992 by CRC Press, Inc.

International Standard Book Number 0-8493-6823-5

Library of Congress Card Number 91-30893

Printed in the United States   1 2 3 4 5 6 7 8 9 0

PREFACE

This volume grew out of a symposium organized by the students of Professor Myron L. Bender, honoring him at his retirement from active teaching from the Department of Chemistry, Northwestern University, Evanston, Illinois. His research focused on the mechanisms of organic reactions, particularly with respect to the understanding of enzymatic catalysis. An overview of his scientific contributions by Fred Menger is included in this volume as is a tribute to him by his friend and colleague, Irving Klotz. Not only was Myron Bender an outstanding research scientist, contributing fundamental discoveries that provide the basis for understanding the mechanisms of enzymatic reactions, but he also was an outstanding teacher. This volume — the contributions of his students — addresses a wide range of bioorganic and enzymatic processes and is a testimony to the continued influence that Professor Bender has had on the development of the field of Bioorganic catalysis.

Valerian T. D'Souza
Joseph Feder

THE EDITORS

Valerian T. D'Souza, Ph.D., presently holds a faculty position in the Department of Chemistry at the University of Missouri-St. Louis. He is involved in developing artificial redox enzymes by covalently attaching flavin to cyclodextrins. These artificial enzymes are expected to have the stability and versatility of cyclodextrins and the catalytic activity of enzymes, making them ideal catalysts for use in industrial environments. They are being patented in collaboration with Millinckrodt Specialty Chemicals Company.

He was born in Kattingere, India and received his Bachelors and Masters degrees from St. Xavier's College in Bombay, India. He received a Ph.D. degree in 1983 based on the work done with H. Harry Szmant at the University of Detroit. He then worked with Myron L. Bender at Northwestern University where he developed a model of Chymotrypsin by covalently attaching imidazole and benzoic acid to cyclodextrins. His research interests are enzyme mechanisms, artificial enzymes, and cyclodextrin chemistry.

Joseph Feder, Ph.D., is President and Chief Executive Officer of Invitron Corporation, St. Louis, Missouri, and Adjunct Professor, Chemistry, University of Missouri-St. Louis.

Dr. Feder received a B.S. degree in biology from Roosevelt University in 1953 and M.S. and Ph.D. degrees from the Illinois Institute of Technology in 1961 and 1964, respectively.

Dr. Feder is a member of the American Chemical Society, the American Society for Biochemistry and Molecular Biology, the American Society for Cell Biology, the American Society for Advancement of Science, the New York Academy of Science, the Tissue Culture Association, and Sigma Xi.

Dr. Feder has authored and coauthored over 120 research articles and abstracts and is the inventor on 33 U.S. patents. He has presented numerous invited lectures and served as symposia chair. His research interests include the biochemical regulation of mammalian cell growth *in vitro,* the development of steady-state continuous mammalian cell culture systems, and the role of glycosylation in enzyme catalysis.

CONTRIBUTORS

Masayasu Akiyama
Department of Applied Chemistry
Tokyo University of Agriculture
  and Technology
Koganei, Tokyo, Japan

Andi O. Chen
Department of Food Science
National Chung Hsing University
Taichung, Taiwan, Republic of
  China

Kenneth A. Connors
Professor
School of Pharmacy
University of Wisconsin
Madison, Wisconsin

Valerian T. D'Souza, Ph.D.
Department of Chemistry
University of Missouri-St. Louis
St. Louis, Missouri

Joseph Feder, Ph.D.
Invitron Corporation
St. Louis, Missouri

Anthony L. Fink
Department of Chemistry
The University of California
Santa Cruz, California

Jeffrey Frye
Division of Natural Sciences
University of Findlay
Findlay, Ohio

Avi Golan-Goldhirsh
The Jacob Blaustein Institute for
  Desert Research
Ben-Gurion University of the
  Negev
Sede Boqer Campus, Israel

Gordon A. Hamilton
Department of Chemistry
Pennsylvania State University
University Park, Pennsylvania

Susan C. Howard
Monsanto Company
St. Louis, Missouri

Ken Ikeda
Faculty of Pharmaceutical
  Sciences
Nagoya City University
Mizuho-ku, Nagoya, Japan

Irving M. Klotz
Department of Chemistry
Northwestern University
Evanston, Illinois

Makoto Komiyama
Department of Industrial
  Chemistry
Faculty of Engineering
University of Tokyo
Hongo, Tokyo, Japan

Yukihisa Kurono
Faculty of Pharmaceutical
  Sciences
Nagoya City University
Mizuho-ku, Nagoya, Japan

W. G. Martin
(Deceased: 01 December 1986)
Institute of Biological Sciences
National Research Council of
  Canada
Ottawa, Ontario, Canada

Fredric M. Menger
Department of Chemistry
Emory University
Atlanta, Georgia

David T. Osuga
Department of Food Science and
 Technology
University of California, Davis
Davis, California

Manfred Philipp
Department of Chemistry
Lehman College and Graduate
 Center
City University of New York
Bronx, New York

Ralph M. Pollack
Laboratory for Chemical
 Dynamics and Department of
 Chemistry and Biochemistry
University of Maryland-Baltimore
 County
Baltimore, Maryland

Galla Rao
Immunicon Corporation
Huntington Valley, Pennsylvania

John F. Sebastian
Department of Chemistry
Miami University
Oxford, Ohio

Frederick C. Wedler
Department of Molecular and Cell
 Biology
Pennsylvania State University
University Park, Pennsylvania

John R. Whitaker
Department of Food Science and
 Technology
University of California, Davis
Davis, California

R. E. Williams
Institute of Biological Sciences
National Research Council
Ottawa, Ontario, Canada

Arthur J. Wittwer
Monsanto Company
St. Louis, Missouri

E. Ziomek
Biotechnology Research Institute
National Research Council of
 Canada
Montreal, Quebec, Canada

This book is dedicated to the memory of

Dr. Myron L. Bender
and
Muriel S. Bender

TABLE OF CONTENTS

A Tribute to Myron L. Bender
Irving M. Klotz

Myron Lee Bender — An Overview of Scientific Contributions
Fredric M. Menger

A TRIBUTE TO MYRON L. BENDER*

Each of us arrives on this earth endowed with some advantages and burdened with some handicaps. And we leave when our infirmities overwhelm us. During the intervening years, each of us has the opportunity to create something for himself, for his family and associates, and for present and future generations.

Myron Bender added far more than one man's portion to the world's fund of scientific knowledge and understanding. Researcher, teacher, writer, he functioned with distinction in all of these activities.

As a researcher he was gifted with a special blend of imagination and disciplined thought that led him to a continual series of fundamental discoveries. The talent was demonstrated in his very first independent paper, in 1951, in which he adapted isotope labels to obtain a fresh view of the mechanistic details of ester hydrolysis. His new insights allowed him to expand his perspective to the broad area of organic hydrolysis reactions, where he soon also discovered the powerful role of imidazole catalysis. In 1960 he published an integrating review of nucleophilic reactions of derivatives of carboxylic acids that became one of the classics of mechanistic organic chemistry. It is one of the most frequently cited papers in the literature of chemistry and was chosen as a citation classic for inclusion in Arnold Thackray's book entitled *Contemporary Classics in Physical, Chemical and Earth Sciences*. In the Soviet Union this paper was translated into Russian and published as a book. Myron frequently complained, good-naturedly, that although this book sold well, it was the only one for which he never received any royalties. I believe this experience colored his views of contemporary socialism, at least as practiced in the Soviet Union.

Bender's vision and sound judgment were manifested again around 1960 when he turned his attention to enzymatic hydrolyses. He had the foresight to recognize that this would become an exciting and challenging field. During the next decade he made remarkable strides in delineating intermediates in the mechanistic pathways of hydrolytic enzymes and in setting kinetic analyses on a sound foundation. His form of analysis set a pattern that was followed by him and others in the subsequent delineation of the molecular mechanisms of action of a wide range of proteolytic enzymes.

During the latter half of his scientific life he opened up new areas of research with enzyme models. He was one of the earliest people to recognize that the binding capacities of the cyclodextrins made them attractive frameworks for the creation of novel catalytic entities, which he called miniature models of enzymes, and he very successfully developed these materials. He also designed even smaller molecules, miniature organic models of enzymes, and overcame formidable synthetic hurdles to generate novel catalysts.

---

* These remarks were made at a memorial service at Northwestern University on October 6, 1988.

The unifying thread throughout Bender's research was hydrolysis reactions. He was recognized internationally at an early age as a pioneer in elucidating the mechanisms of these reactions, particularly with enzymes. Many honors were bestowed upon him, including election to the National Academy of Sciences, the American Chemical Society Midwest Award, and an honorary D.Sc. degree from his alma mater, Purdue. He was continually chosen as a plenary or major lecturer at national and international meetings.

Many students and postdoctorate fellows were attracted to his laboratories. He had the gratification of seeing so many of them progress to distinguished careers of their own in academic or industrial institutions. Among the many academics one thinks of Brubacher (Waterloo), Connors (Wisconsin), Fink (Santa Cruz), Hamilton (Penn State), Kaiser (Rockefeller), Kezdy (Chicago), Means (Ohio State), Menger (Emory), Pollack (Maryland), Silver (Amherst), Van Etten (Purdue), Wedler (Penn State), Whitaker (Davis), Williams (Kent, England), Zerner (Queensland, Australia). Some of his particularly successful industrial people include Feder (Monsanto), Turnquest (Arco), Valenzuela (Chiron); and I am sure I have overlooked many.

Bender was also intimately involved in teaching undergraduates and graduate students at all levels. At the time of his arrival at Northwestern, the new honors premedical program had just been inaugurated, and he took the responsibility of developing and teaching a special course in organic chemistry for these gifted and knowledgeable young people. Those original classes generated many exceptionally fine physicians who have now been in practice for twenty years. Whenever you encounter them, they refer with affection and admiration to Professor Bender.

His books have also made the basic principles and the subtleties of kinetics and catalysis intelligible and exciting to generations of students throughout the world. Translations have been printed in Spanish, Japanese, and German. It is ironic that essentially all of these books were written after he had been elected to the National Academy of Sciences. He was in the midst of writing still another during the past summer.

For almost three decades, from the year of his arrival at Northwestern, Myron was a member of the post-luncheon walking club led by Professor Pines, which makes a daily inspection tour from the north end to the south end of campus and back. During these walks Myron contributed cogently to our analyses of pressing world problems and abstract philosophical issues, and he also gave us periodic reports on the progress of each of his three sons, Alec, Bruce, and Steve, of whom he was so proud. Twice a year, at the south end of our route he would detach himself unobtrusively from the group and disappear quietly. After many years, we could no longer contain our curiosity as to where he went on these occasions, and Professor Pines, our doyen, was commissioned to ask him. Myron informed us that on each occasion, either Muriel's birthday was the following day or Mother's Day was the following Sunday and that each time he had gone into town to buy Muriel a suitable gift. It is a doubly poignant tragedy that Muriel is not here today to hear these testimonies to her lifelong companion.

No human life is free of misfortune. Myron Bender carried more than one person's burden of afflictions. But he also showed more than one person's stock of courage and inner strength of spirit. To the very end, he continued to learn and teach, regardless of impediments.

The congregation assembled at this memorial service is an eloquent testimony to his achievements. His memory will be cherished by his many students, colleagues, and friends.

Irving M. Klotz

Myron Lee Bender — An Overview of Scientific Contributions*
(May 20, 1924 - July 29, 1988)

How does one summarize a lifetime of dedication and excellence? I might begin with biographical data and then relate more specifically the true impact that Myron Bender has had on modern chemistry.

Myron Lee Bender was born and raised in St. Louis, Missouri, and obtained both B.S. (1944) and Ph.D. degrees (1948) from Purdue University (his Ph.D. thesis research was performed under the direction of Henry B. Hass). Following a postdoctoral research year with Paul D. Bartlett at Harvard University, a second year as an AEC fellow in Frank H. Westheimer's laboratory at the University of Chicago, and a year on the faculty at the University of Connecticut, he accepted in 1951 a position in the Department of Chemistry of the Illinois Institute of Technology and stayed there for nine years. In 1960, he moved to the Department of Chemistry at Northwestern University, where he remained throughout the rest of his career. During his career, he produced over 200 publications, 18 monographs, and 5 books. He received the Midwest Award of the American Chemical Society, a distinguished Fulbright Fellowship, an Honorary Degree from Purdue University, and membership in the National Academy of Sciences.

But information of this sort does not adequately describe the real impact that Myron's research program has had on organic chemistry. It would be far better in this regard to focus specifically on how his work has contributed to the development of chemical thought. This is not an easy task in a short presentation. Since Myron's work has always been multifaceted (crossing the boundaries of physical organic chemistry, bioorganic chemistry, enzymology, inorganic chemistry, and colloid chemistry), I must necessarily be selective in delineating below the highlights of his scientific career. In the process of choosing among his accomplishments, I experienced the joy of rereading many of his papers. One cannot help being struck by their uniform clarity and brilliance — no convoluted arguments, no overinterpreted data, no handwaving, no unnecessary complexities to impress the reader — just gems of inductive reasoning. Cited below are a few articles I admire most.

1. **M. L. Bender,** Oxygen exchange as evidence for the existence of an intermediate in ester hydrolysis, *J. Am. Chem. Soc.,* 73, 1626, 1951.

This paper, the first of a family of papers dealing with ester and amide hydrolyses, showed that an ester labeled with ^{18}O at the carbonyl exchanges its ^{18}O with solvent during hydrolysis; a tetrahedral intermediate is thereby strongly indicated. The "^{18}O" papers launched Myron into national prominence; they also were instrumental in ushering in the era of "aqueous" physical organic chemistry which was to thrive for three decades thereafter.

* From *Bioorganic Chemistry,* 17, 252, 1989. With permission.

Myron was one of the leading physical organic chemists who did for reactions in water what Winstein and others did for reactions in acetic acid. The mechanistic detail ultimately achieved by Myron was astounding. For example, the ^{18}O exchange work of Bender and Heck in 1967 led to the proposal of a *general base*-catalyzed attack of water on an ester carbonyl forming a tetrahedral intermediate which decomposes *spontaneously* to product, but reverts back to reactants by *acid catalysis* (Equation 1). Three steps and three different modes of reactivity! It was also in this paper that Myron proposed that proton transfer within a tetrahedral intermediate could be rate-limiting — an idea far ahead of its time, as it turned out.

$$CF_3COSEt + H_2O \underset{k_2(H_3O^+)}{\overset{k_1(H_2O)}{\rightleftharpoons}} [I] \overset{k_3}{\longrightarrow} CF_3COOH \qquad (1)$$

2. **M. L. Bender, F. J. Kezdy, and B. Zerner,** Intramolecular catalysis in the hydrolysis of p-nitrophenyl salicylate, *J. Am. Chem. Soc.*, 85, 3017, 1963.

During the time that West Coast groups were developing the chemistry of "anchimeric assistance" or "neighboring group participation" in acetic acid, the bioorganic component of physical organic chemistry was doing the same with "intramolecular catalysis" in water. Myron was, undoubtedly, a key figure in the development of our present understanding of intramolecularity, one of the most importance concepts in organic chemistry. The above paper, providing the first proven case of an intramolecular general base catalysis, illustrates the point. The paper also serves to illustrate a classic piece of physical organic reasoning. Myron was faced with the problem of differentiating between a general base mechanism (Structure 1) and a kinetically equivalent hydroxide/general acid mechanism (Structure 2). He solved this problem by showing that anionic nucleophiles (e.g., azide ion) do not manifest intramolecular catalysis, thus ruling out mechanism 2 that had *a priori* been favored by others. This typifies the simple, quiet beauty of Myron's work.

1   **2**

3. **M. L. Bender, F. J. Kezdy, and C. R. Gunter,** The anatomy of an enzymatic catalysis: α-chymotrypsin, *J. Am. Chem. Soc.*, 86, 3714, 1964.

Myron's research over the years was concerned with the mechanism of organic reactions (particularly those of enzyme "models") along with the mechanism of the enzymatic process themselves. There have been others who confined themselves to one aspect or the other, but Myron was deeply involved in both. This probably accounted in large measure for his success, because the riddle of why enzymes react so fast can be solved only by interrelating the physical organic chemistry with enzymology. The above classic article by Bender and co-workers represents, in the opinion of many, the most penetrating analysis of an enzyme mechanism available at that time. Naturally, Myron drew on the work of the other laboratories in developing the chymostrypsin mechanism which is still valid today. But many of the ideas originated from his own laboratory. Thus, he was the first to provide spectrophotometric evidence for the acyl-enzyme intermediate. He showed how amides and esters have different rate-determining steps. He taught the world how to titrate enzymes and thus place enzyme kinetics on a firm, quantitative footing. He synthesized a new enzyme, thiosubtilism, in which a serine hydroxyl at the active site was replaced by a thiol group. Charge relay mechanisms, rate-determining binding, specific binding of water, enzyme dimerization, aging processes, cannibalistic denaturation — all came under his scrutiny. Chymotrypsin, trypsin, papain, elastase, and acetylcholinesterase yielded their secrets to Myron's group. Can there be any doubt that there was something uniquely successful about Myron's approach to the enzyme problem?

4. **R. C. VanEtten, J. F. Sebastian, G. A. Clowes, and M. L. Bender,** Acceleration of phenyl ester cleavage by cycloamyloses: a model for enzymatic specificity, *J. Am. Chem. Soc.,* 89, 3242, 1967.

Myron was among the very first to recognize that the chemistry of the future would involve complexation (perhaps stereospecific complexation) of reactants prior to the actual reaction. This is, after all, the way enzymes function. In order to mimic such enzymatic behavior, Myron used cyclodextrin systems which complex and then react with small organic substrates. His work in the area is described in a series of articles and a book, of which the above-cited paper is an example. Its import can be appreciated from two sentences taken from the abstract: "The cycloamyloses cause a markedly stereoselective acceleration of phenol release from a variety of substituted phenyl acetates in alkaline solution." "The reaction system constitutes a striking model for the lock and key theory of enzymatic specificity proposed by Emil Fischer." In my opinion, the Bender paper is more than just an interesting enzyme model. It initiated the era of "biomimetic chemistry". It paved the way for others (particularly Breslow, Cram, Lehn, Murakami, and Tabushi) to experiment with preassociative mechanisms. If Myron had accomplished nothing else other than this cyclodextrin work, he would have still left a considerable mark on physical organic chemistry.

5. I. M. Mallic, V. T. D'Souza, M. Yamaguchi, J. Lee, P. Chalabi, R. C. Gadwood, and M. L. Bender, An organic chemical model of the acyl-α-chymotrypsin intermediate, *J. Am. Chem. Soc.*, 106, 7252, 1984.

Dr. Bender's group reported the synthesis of what is certainly one of the most sophisticated chemical models of the acyl-enzyme to date. It incorporates all three components believed present at the active site of the enzyme: an imidazole ring, a carboxylate, and a hydroxyl (the latter of which becomes acylated during the reaction between the enzyme and an ester substrate). The Bender model (see compound 3) has a hydrolysis rate approximately equivalent to the actual acylenzyme intermediate and 154,000 times faster than an ordinary ester. Synthetic organic chemistry, physical organic chemistry, and biological relevance are combined here in a magnificent system.

Myron Bender's science will not fade away. Obviously, the principles of nature that he uncovered during his career will be incorporated into the chemical literature for a long time to come. But, more than that, I have noticed that Myron's style of science pervades the publications of myself and all the other of his disciples. And I have noticed that our students, Myron's scientific "grandchildren", also possess the Bender aura, sometimes without knowing it. The debt here is multigenerational and eternal.

Myron had a wonderful wife, Muriel, whose love was apparent whenever they were seen together. This love, I am certain, enabled Myron to cope for many years with the effect of a terrible stroke, and to continue to produce good chemistry despite it, although a similar affliction would have all but destroyed a man not as blessed. One wonders if greatness in science is ever possible without the force of mutual affection. Muriel passed away soon after Myron as if Myron had called her home.

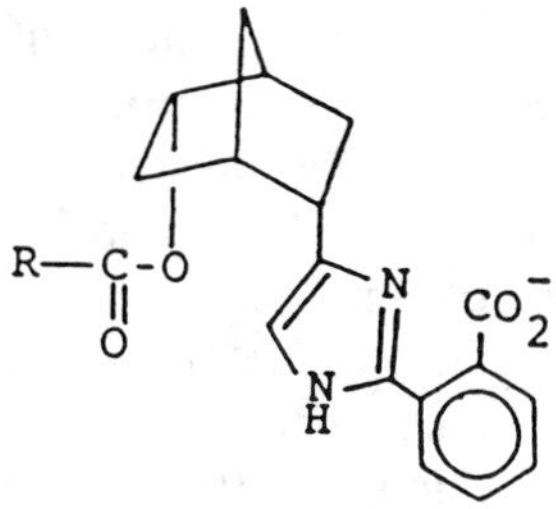

3

Fredric M. Menger
Emory University
Atlanta, Georgia

Chapter 1

CATALYTIC ACTIVITY OF 2-SUBSTITUTED IMIDAZOLES FOR HYDROLYSIS OF ACYL ESTER DERIVATIVES

Masayasu Akiyama

TABLE OF CONTENTS

I. INTRODUCTION

In 1968, as a postdoctoral fellow with Professor Bender, I wished to study on an advanced serine-protease model which would be an extension of the previous searching investigation.[1] It seemed to be appropriate to involve a base such as an imidazolyl group for the facilitation of cyclodextrin-esterolytic reactions. The recent report by D'Souza and Bender has described the participation of imidazolyl and carboxyl groups.[2]

In serine proteases, a histidine imidazolyl group is believed to accept a proton from a serine hydroxy group and to give the proton to the leaving amine moiety of a substrate to form an acyl-enzyme intermediate.[3] In the papain-catalyzed hydrolysis of an amide substrate, a rate-determining proton transfer from a protonated imidazolyl group to a tetrahedral intermediate is suggested,[4] and in the action of ribonuclease, a protonated imidazolyl group is thought to act concertedly on a substrate with another nearby imidazolyl group.[5]

Although preliminary experiments at that time suggested that imidazole might act as a general base for cyclodextrin cleavage of an aryl carboxylate, it was difficult to assess such a general base catalysis contribution from a rate retardation caused by cyclodextrin addition. Imidazole is an excellent nucleophile[6] and has a simple structure, when compared with a multiregulating enzyme system. Therefore, it becomes necessary to change the steric environment of the molecule in order to provide systems where even good leaving aryl esters can be subjected to imidazole general base catalysis.

After returning to Tokyo, I decided to initiate a systematic study on imidazole catalysis for hydrolysis of acyl-ester derivatives, using a series of 2-substituted imidazoles. The 2-position of the imidazole ring is a suitable site for substitution because the 2-substituent can affect both nitrogen atoms equally and effectively, irrespective of the imidazole prototropy.

Our attention is given particularly to the following points.

1. By how much can the nucleophilic reactivity of imidazole bases be diminished by the presence of steric bulk in the 2-position?
2. Does the mechanism of imidazole catalysis change from nucleophilic to general base as the steric bulk increases?
3. Does a hydroxy group proximate to the imidazole function facilitate general base catalysis when the nucleophilic reactivity is effectively reduced?
4. By how much does the steric bulk of the imidazole ring diminish the general base activity and does a hydroxy group in the substituent facilitate general base catalysis?
5. Does a substituent on an imidazolium ion affect general acid catalytic activity?

The acyl-ester derivatives used as substrates were 4-nitrophenyl acetate for nucleophilic catalysis,[7] ethyl dichloroacetate for general base catalysis,[8] and diethyl phenyl orthoformate for general acid catalysis.[9]

II. NUCLEOPHILIC CATALYSIS

The imidazole-catalyzed hydrolysis of 4-nitrophenyl acetate, which is known for its susceptibility to imidazole nucleophilic catalysis,[6] was carried out in water by utilizing a series of 2-substituted imidazoles as buffer under pseudo first-order reaction conditions.[7] The imidazole base catalysis may be expressed by Equation 1, where k_0 is a catalytic constant due to the solvent species.

$$k_{obs} = k_{im}[Im]_f + k_0 \qquad (1)$$

The values of k_{im} in water at 30°C (and at two additional temperatures for some of the imidazoles, though not given here) are listed in Table 1 together with the pK_a values at 30°C. The deuterium oxide solvent isotope effect was determined for some of the imidazoles. Activation parameters were calculated from the data obtained. These results are also collected in Table 1.

Because the pK_a values of the imidazoles are different, direct comparison of the k_{im} values is not useful. Therefore, an extrapolated rate value may be derived for each imidazole from its k_{im}. Two assumptions are made for this purpose: (1) each substituent series obeys the Brönsted catalysis law and (2) there is an imidazole with a hypothetical pK_a of 7.80 in the individual series. For the Brönsted plot, a universal slope of 0.80 was used, because the catalysis is considered to be nucleophilic. Parallel plots of the log k_{im} vs. pK_a were made as shown in Reference 7, and the extrapolated rate value (R_x) for each imidazole was given by taking the k_{im} value at pK_a 7.80. In the nucleophilic reactions of 4-nitrophenyl acetate, Bruice and Lapinski showed parallel Brönsted correlations with a common slope of 0.80 for different types of bases.[10] Jencks and Gilchrist also showed correlations with slope 0.80 for similar reaction series which are comprised of structurally related amine nucleophiles.[11]

A measure of the steric effect of a substituent in the imidazole base is estimated by Equation 2, where R_H represents the extrapolated rate value of the parent imidazole.

$$\text{measure of steric effect} = \log(R_X/R_H) \qquad (2)$$

Table 1 also contains values of R_x and $\log(R_x/R_H)$. A plot of log (R_X/R_H) values vs. the Taft steric constant parameters[12] was done as shown in Reference 7. The correlation was rather good except for the t-butyl group. Hydroxyalkyl groups were tentatively plotted on the same scale as the corresponding alkyl groups, without respect to their oxygen demand. The sensitivity

TABLE 1

Substituted Imidazole-Catalyzed Hydrolysis of 4-Nitrophenyl Acetate in Water[a]

Imidazole substituent	pK_a[b]	Second-order rate constant (M^{-1} min^{-1})		$-\log(R_x/R_H)$	k_{im}^H/k_{im}^D	$\Delta H^\ddagger$ (kcal/mol)	$\Delta S^\ddagger$ (cal/mol · K)
		$10^3 k_{im}$ (30°C)	$10^3 R_x$[c]				
2-H	7.19	800	2500	0		8.7	30
2-HOCH$_2$	6.90	25	130	1.3	1.2	9.7	34
2-Me	8.10	160	94	1.4		8.6	34
2-Et	8.07	60	37	1.8			
2-HOCH (Me)CH$_2$	7.68	23	29	1.9			
2-Pr^i	8.01	13	9.1	2.4	1.0	10.6	32
2,4,5-Me$_3$	9.03	7.3	0.76	3.5	1.6	11.0	29
2-HOCH$_2$C (Me)$_2$	7.54	0.049	0.079	4.5	1.7	10.3	44
2-Bu^t	7.93	0.03	0.023	5			

a   Acetonitrile (0.8% v/v) and ionic strength 1.0 (KCl); 4-nitrophenyl acetate 1.0×10^{-4} M

b   At 30°C and ionic strength 1.0

c   Extrapolated rate value.

From Akiyama, M., Hara, Y., and Tanabe, M., *J. Chem. Soc. Perkin Trans. 2*, 288, 1978. With permission.

TABLE 2
Substituted Imidazole-Catalyzed Hydrolysis of Ethyl Dichloroacetate in Water[a]

Imidazole substituent	pK_a^b	Second-order rate constant $(M^{-1} min^{-1})$ $10^2 k_{im}$ (30°C)	$10^2 R_x^c$	$-\log (R_x/R_H)$	$\Delta H^{\ddagger}$ (kcal/mol)	$\Delta S^{\ddagger}$ (cal/mol · K)
2-H	7.11	9.4	20	0	9.1	41
2-HOCH$_2$	6.73	5.7	18	0.034		
2-Me	8.10	22	16	0.090	7.8	44
2-HOCH(Me)CH$_2$	7.53	9.3	13	0.20	8.7	45
2-Et	7.94	10	8.7	0.36		
2-Pr^i	7.90	8.0	7.2	0.44	7.5	47
2-HOCH$_2$C(Me)$_2$	7.58	4.6	5.9	0.53	9.6	41

[a] Acetonitrile (0.5% v/v) and ionic strength 1.0 (KCl); ester concentration $3.0 \times 10^{-3} M$.
[b] At 30°C and ionic strength 1.0
[c] Extrapolated rate value.

From Akiyama, M., Ikjima, M., and Hara, Y., *J. Chem. Soc. Perkin Trans 2*, 1512, 1979. With permission.

to the Taft constant of the present reaction was 1.33 from the slope of the plot, although the *t*-butyl derivatives deviated considerably.

Both the solvent isotope effect ($k_{im}H/k_{im}D = 1.7$) and large negative entropy of activation for 2-(1,1-dimethyl-2-hydroxyethyl)imidazole suggest a mechanism different from nucleophilic catalysis.

A systematic change of the substituent in the 2-position of imidazole leads to a gradual decrease in the nucleophilic reactivity, approaching almost negligible activity for the *t*-butyl group. It might be expected that reduction of nucleophilic reactivity by 10^4 to 10^5 would result in a mechanistic changeover from nucleophilic to general base catalysis. However, when steric bulk is present, general base activity is also decreased, though moderately. The combined effects of the rate retardation by steric bulk and the slight acceleration by a hydroxy group may be a partial occurrence of the general base, as observed with 2-(1,1-dimethyl-2-hydroxyethyl)imidazole.

III. GENERAL BASE CATALYSIS

The hydrolysis of ethyl dichloroacetate, which is known for its susceptibility to imidazole general base catalysis,[13] was carried out in imidazole buffer at constant temperature[8] by estimating the residual ester by a hydroxylamine assay.[14]

Plots of $(k_{obs} - k_0)/[Im]_t$ vs. the fraction of free-base concentration showed good linear relations, with a zero intercept. The second-order rate constants (k_{im}) for substituted imidazoles are summarized in Table 2 together with the pK$_a$ values. It was difficult to include the data for 2-*t*-butylimidazole because of its limited solubility.

Activation parameters (Table 2) show enthalpies of reasonable magnitudes and large negative entropies, the latter being characteristic of a termolecular transition state where a molecule of water is involved in addition to the catalyst and substrate molecules. The k$_{im}$ values for the substituted imidazoles at 30°C are not in the order expected from the size of the substituents. In order to evaluate the steric effect of the substituents, it is necessary to compare these rate constants with a standard. This may be done similarly by assigning an extrapolated rate value (R$_X$) to each imidazole. In the hydrolysis of ethyl dichloroacetate, Jencks and Carriuolo have shown a slope of 0.47 for a Brönsted plot for a series of relatively unhindered bases, including imidazole.[13] A measure of the steric effect of a substituent in the imidazole is given by Equation 2. Values of R$_X$ and log(R$_X$/R$_H$) are also included in Table 2. Plots of log(R$_X$/R$_H$) vs. the Taft steric constant parameters were made similarly as shown in Reference 8.

Two parallel lines with a slope of 0.26 were drawn, an upper line representing the hydroxyalkyl groups and the bottom one the alkyl groups. The steric parameter sensitivity of 0.26 for these substituents makes a striking contrast with that of 1.33 for nucleophilic catalysis by the same imidazoles. The hydroxy group favors catalysis despite the larger steric demand than that for hydrogen. From the difference in the two parallel lines, a rate increase factor of 1.7 was calculated. It must be remembered that no allowance for the steric effect of the oxygen atom was made in that treatment. The rate enhancements for ester hydrolysis by a neighboring hydroxy group have been considered in terms of hydrogen bonding, acid-base catalysis, nucleophilic reactions, and solvent sorting or microscopic solvent effects.[15] The present case is explained most plausibly by this last effect.

If the imidazoles participate solely as a proton base in the general catalysis, there would be no steric effect by the substituents. Both a small but sizable steric effect and a hydroxy group effect indicate that the substituent in the imidazole base is involved to some extent in the transition state of the catalyzed reaction. It is known that there are a few hydrophilic functions along the periphery of enzyme active sites.[3] The role of the hydroxy group in the imidazoles may be related to such functions.

IV. GENERAL ACID CATALYSIS

Diethyl phenyl orthoformate was chosen as substrate[9] because it has been shown to be hydrolyzed through general acid catalysis by a series of carboxylic acids.[16] The partial resemblance of an orthoester to a tetrahedral intermediate makes this hydrolysis a model reaction for the breakdown of an enzyme tetrahedral intermediate.

An increase in slope with decreasing pH was taken as evidence for general acid catalysis by imidazolium ions. Thus, Equation 3 holds.

TABLE 3

Rate Constants for Hydrolysis of Diethylphenyl Orthoformate Catalyzed by 2-Substituted Imidazoles in 50% Dioxane-Water (v/v)[a]

Imidazole substituent	pK_a^b	$10^2 k_{imH}^+/M^{-1}\,s^{-1}$			$10^2 R_X^c/M^{-1}\,s^{-1}$	$-\log (R_X/R_H)$
		30°C	35°C	40°C		
2-H (Dioxane-D$_2$O)	6.40	7.74	10.8	15.6	4.07	0
		3.74				
2-Me	7.34	3.05			4.37	-0.03
2-Et	7.31	3.13			4.37	-0.03
2-HOCH(Me)CH$_2$	7.06	3.25			3.47	0.07
2-Pr^i	7.20	2.56	4.00	6.11	3.16	0.11
2-Bu^t	7.10	1.97			2.19	0.27
2-HOCH$_2$C(Me)$_2$	6.99	1.64	2.68	3.80	1.66	0.39

[a] Ionic strength 0.1 (KCl); diethylphenyl orthoformate 1×10^{-4} mol l^{-1}.
[b] Taken as the pH of the half-neutralized buffer solution in 50% dioxane-water at 30°C and ionic strength 0.1.
[c] Extrapolated rate value.

From Ihjima, M., Fukuyama, M., Kobayshi, T., Hirakawa, T., and Akiyama, M., *J. Chem. Soc. Perkins Trans. 2*, 669, 1987. With permission.

$$k_{obs} = k_0 + k_{imH^+}[ImH^+] \qquad (3)$$

The value obtained at zero buffer concentration is the rate constant due to spontaneous hydrolysis, k_0, by water and hydronium ion. The values of k_{imH^+} for 1:1 buffer at 30°C and at additional temperatures for some of the imidazoles are given in Table 3 together with the pK_a values under experimental conditions.

Scheme 1 represents the accepted structure of the transition state of the acid-catalyzed hydrolysis of orthoesters. To confirm general acid catalysis by imidazolium ions, $(k_{obs} - k_0)/[Im]_{total}$ was plotted against the fraction of the acidic component of the buffer, and straight lines with a zero intercept were obtained.

The second-order rate constants for 2-substituted imidazoles at 30°C decrease with increasing bulkiness of the substituents (Table 3). The value of k_{imH^+} for 2-t-butylimidazole, which has the most bulky substituent in the series, is only five times as small as that for the parent imidazole. It can also be said that steric hindrance is less important in general acid catalysis.

To evaluate the steric effect of the substituents, $\log(k_{imH^+}/k_{H\,imH^+})$, where $k_{H\,imH^+}$ is the second-order rate constant for the parent imidazole, was plotted against the Taft steric constant parameter, E_s^{12} (not shown here). Hydroxyalkyl groups were tentatively plotted on the same scale as the corresponding alkyl groups, although the oxygen atom has its own bulk. The slope of the plot represented the susceptibility of the steric effects on the reaction and it was estimated to be $0.22 + 0.03$ (standard error).

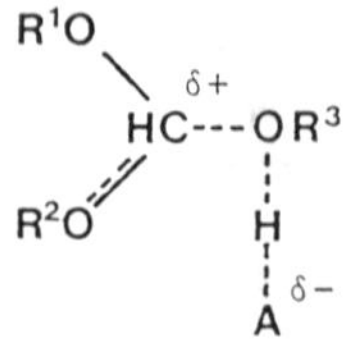

$$R^1O \diagdown \quad \underset{HC}{\overset{\delta+}{\cdots}} OR^3$$

R²O, H, A, δ−

SCHEME 1.   General acid catalysis usually shown without solvent molecules. (From Ihjima, M., Fukugama, M., Kobayshi, T., Hirakawa, T., and Akiyama, M., *J. Chem. Soc. Perkin Trans. 2*, 669, 1987. With permission.)

EtO, EtO, HC, O, H, O, H, H, N, NH, R, δ+, δ+

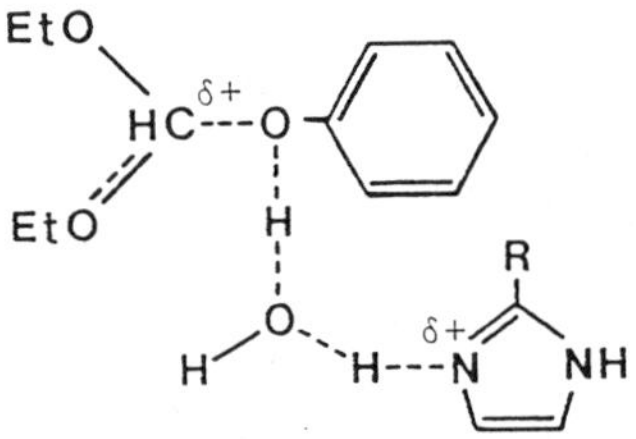

SCHEME 2.   General acid catalysis properly shown with a water molecule. (From Ihjima, M., Fukuyama, M., Kobayshi, T., Hirakawa, T., and Akiyama, M., *J. Chem. Soc. Perkin Trans. 2*, 669, 1987. With permission.)

The small susceptibility implies that the rate-determining proton transfer from an acid to the substrate is indirect. A mechanism represented by Scheme 2 shows that at least one solvent water molecule intervenes between an imidazolium ion and the orthoester, stabilizes the transition state by separating the charge-developing species, and facilitates proton transfer through the hydrogen bond formed. In fact, many proton-transfer reactions in hydrolytic solvents have been known to proceed with participation of solvent water molecules.[17] General acid catalysis is depicted without water molecules, as in Scheme 1. However, it should correctly be shown as in Scheme 2.

While general acid catalysis requires at least one solvent molecule, what is the requirement for general base catalysis? According to the principle of microscopic reversibility of a chemical reaction, general base catalysis in one direction should be general acid catalysis in the opposite direction. Insofar as general acid catalysis entails a water molecule, general base catalysis also requires one water molecule as discussed by Ihjima et al.[9]

V. GENERAL BASE CATALYSIS FOR CYCLODEXTRIN-ESTEROLYTIC REACTIONS

Cyclodextrins have been utilized as enzyme models.[18] Esterolytic reactions catalyzed by cyclodextrins proceed through nucleophilic attack by their alkoxide

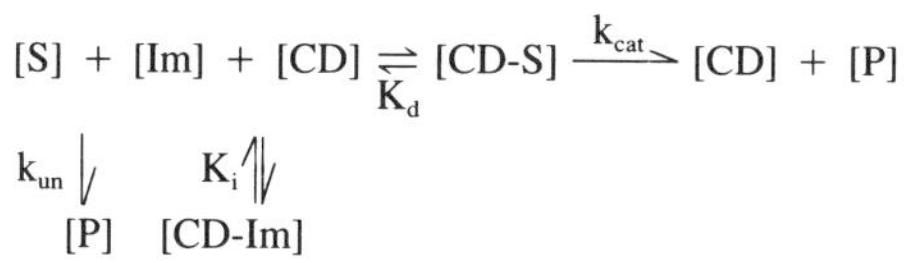

$$[S] + [Im] + [CD] \underset{K_d}{\rightleftharpoons} [CD\text{-}S] \xrightarrow{k_{cat}} [CD] + [P]$$

$$k_{un} \Big\downarrow \qquad K_i' \Big\Vert$$

$$[P] \qquad [CD\text{-}Im]$$

SCHEME 3. The reaction sequence for hydrolysis of 3-nitrophenyl acetate in the presence of β-cyclodextrin and an imidazole base. (From Akiyama, M., Ohmachi, T., and Ihjima, M., *J. Chem. Soc. Perkin Trans. 2*, 1511, 1982. With permission.)

ions, and the reactions are highly dependent on the concentration of hydroxide ion in solution.[1] In view of discussions about the charge relay system for enzymes,[3] it was desirable to construct a model which works by a general base mechanism through a covalently attached base. Before synthesizing appropriate models, it was useful to examine the effects of added bases on cyclodextrin reactions.

The hydrolysis of 3-nitrophenyl acetate was carried out in the absence or presence of β-cyclodextrin with substituted imidazoles as buffer at 30°C and ionic strength 0.20.[19] The initial concentrations of the imidazole [Im], the cyclodextrin [CD], and the ester [S] were on the order of 10^{-1}, 10^{-3}, and 10^{-4} M, respectively.

The reaction sequence for the initial stage of the hydrolysis is shown in Scheme 3. Here, the substrate is converted into products via two pathways represented by the first-order rate constants, k_{un} and k_{cat}. The overall rate is given by Equation 4, where Equation 5 holds. From Equations 4 and 5, Equation 6 is obtained.

$$\text{Rate} = k_{obs}[S]_0 = k_{un}[S] + k_{cat}[CD\text{-}S] \tag{4}$$

$$[S]_0 = [S] + [CD\text{-}S] \tag{5}$$

$$(k_{obs} - k_{un})[S]_0 = (k_{cat} - k_{un})[CD\text{-}S] \tag{6}$$

The k_{un} pathway is more precisely represented by Equation 7.

$$k_{un} = k_0 + k_{Im}[Im]_t \tag{7}$$

The k_{cat} pathway is expressed by Equation 8 if the imidazole used as buffer has catalytic activity.

$$k_{cat} = k_{cat\text{-}OH}[OH] + k_{cat\text{-}Im}[Im]_f \tag{8}$$

Both $k_{cat\text{-}OH}$ and $k_{cat\text{-}Im}$ are the second-order rate constants for CD-S due to hydroxide ion and the imidazole base, respectively. At a given constant pH, $k_{cat\text{-}OH}[OH]$ is constant. Combination and rearrangement of equations give

Equation 9, where the apparent dissociation constant, K_{app}, is defined by Equation 10.

$$\frac{1}{k_{obs} - k_{un}} = \frac{K_d(K_i + [Im]_0)}{[CD]_0 K_i (k_{cat} - k_{un})} + \frac{1}{k_{cat} - k_{un}} \qquad (9)$$

$$K_{app} = \frac{K_d(K_i + [Im]_0)}{K_i} = K_d + \frac{K_d[Im]_0}{K_i} \qquad (10)$$

Thus, the observed rate constants (k_{obs}) were plotted in terms of a Lineweaver-Burk-type equation (9) and the values of k_{cat} and K_{app} were obtained. Plots of k_{cat} and of k_{un} vs. $[Im]_f$ were made for some of the imidazoles. The slope and intercept of the k_{cat} plot yielded the values of k_{cat-Im} and $k_{cat-OH}[OH]$, and, similarly, the k_{un} plot gave k_{Im} and k_0, respectively, as described in Equations 7 and 8. The various constants obtained are summarized in Table 4. A large increase in k_{cat} relative to k_{un} or a larger k_{cat-Im} than k_{Im} in Table 4 for the cases of 2-isopropyl-, 2-(1,1-dimethyl-2-hydroxyethyl)-, and 2,4,5-trimethyl-imidazole clearly favors situation in an illustration (Scheme 4). A Brönsted plot of $\log(k_{cat-Im})$ vs. pK_a for the imidazoles yielded a straight line with a slope of 0.62, as shown in Reference 19. Because the nucleophilic activity of substituted imidazoles is much more affected by their substituents than is general base activity[7,8] and k_{Im} and k_{cat-Im} are quite different functions of pK_a, a value of 0.62 is a good indication of general base assistance. Solvent D_2O isotope effects obtained for 2-isopropyl- and 2,4,5-trimethyl-imidazole (Table 4) are also in line with this result. The values of about 2 for k_{cat-Im} suggest general base catalysis, whereas those of about 1 for k_{Im} indicate nucleophilic catalysis. Our present results favor general base catalysis by an added imidazole base in cyclodextrin-aryl ester cleavage.

In conclusion, perturbation of imidazole-catalyzed reactions by introduction of steric bulk at the 2-position of the imidazole ring reveals several interesting aspects of the reactions, which lead to a better understanding of imidazole catalysis. Work with 2-substituted imidazoles continues in other laboratories.[20]

ACKNOWLEDGMENT

The author thanks his colleagues, whose names appear in the references, for their sincere collaboration.

TABLE 4
Various Constants for Hydrolysis of 3-Nitrophenyl Acetate by β-Cyclodextrin in Substituted-Imidazole Buffers

Imidazole substituent	pK_a[a]	$10^3 k_0/s^{-1}$	$10^3 k_{Im}/M^{-1}$ s^{-1}	$10^3[OH^-]$ $k_{cat\text{-}OH}/s^{-1}$	$10^3 k_{cat\text{-}Im}/M^{-1}$ s^{-1}	$10^3 K_d/M$	K_I/M
2-H	7.19	2.4	233	0.75	2.5	9.4	0.15
2-Me	8.10	0.22	36.6	7.4	16	7.3	0.51
2-Pr^i (H$_2$O)	8.01	0.16	2.8	5.1	8.4	4.1	0.081
(D$_2$O)		0.14	2.1	3.7	4.6	5.2	0.20
2-HOCH$_2$C(Me)$_2$	7.54	0.55	0.066	2.1	4.3	6.8	0.058
2,4,5-Me$_3$ (H$_2$O)	9.03	0.45	0.75	35.3	35.5	5.5	0.14
(D$_2$O)		0.27	0.89	26.8	16.0	4.7	0.17

[a] These values taken from Reference 7.

From Akiyama, M., Ohmachi, T., and Ihjima, M., *J. Chem. Soc. Perkin Trans. 2*, 1511, 1982. With permission.

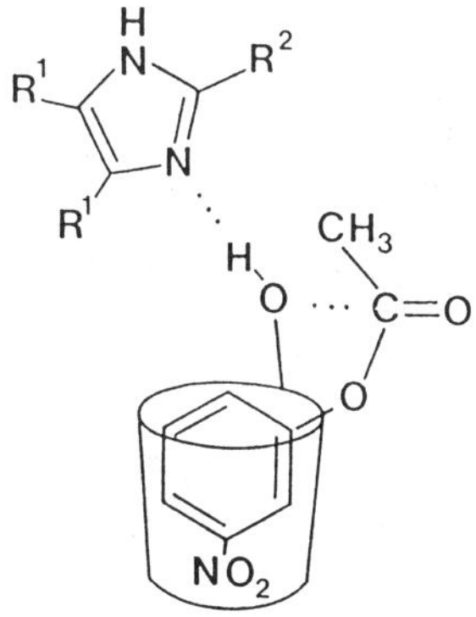

SCHEME 4. Schematic illustration of the hydrolysis of 3-nitrophenyl acetate with substituted imidazoles in the presence of β-cyclodextrin. (From Akiyama, M., Ohmachi, T., and Ihjima, M., *J. Chem. Soc. Perkin Trans. 2*, 1511, 1982. With permission.)

REFERENCES

1. **VanEtten, R. L., Sebastian, J. F., Clowes, G. A., and Bender, M. L.**, *J. Am. Chem. Soc.*, 89, 3242, 1967.
2. **D'Souza, V. T. and Bender, M. L.**, *Acc. Chem. Res.*, 20, 146, 1982.
3. **Blow, D. M.**, *Acc. Chem. Res.*, 9, 145, 1976.
4. **Lowe, G. and Yuthavong, Y.**, *Biochem. J.*, 124, 117, 1971.
5. **Findlay, D., Herrises, D. G., Mathias, A. P., Rabin, B. R., and Ross, C. A.**, *Nature*, 190, 781, 1961.
6. **Bender, M. L. and Turnquest, B. W.**, *J. Am. Chem. Soc.*, 79, 1652, 1957.
7. **Akiyama, M., Hara, Y., and Tanabe, M.**, *J. Chem. Soc. Perkin Trans. 2*, 288, 1978.
8. **Akiyama, M., Ihjima, M., and Hara, Y.**, *J. Chem. Soc. Perkin Trans. 2*, 1512, 1979.
9. **Ihjima, M., Fukuyama, M., Kobayshi, T., Hirakawa, T., and Akiyama, M.**, *J. Chem. Soc. Perkin Trans. 2*, 669, 1987.
10. **Bruice, T. C. and Lapinski, R.**, *J. Am. Chem. Soc.*, 80, 2265, 1958.
11. **Jencks, W. P. and Gilchrist, M.**, *J. Am. Chem. Soc.*, 90, 2622, 1968.
12. **Taft, R. W.**, in *Steric Effects in Organic Chemistry*, Newman, M. S., Ed., John Wiley & Sons, New York, 1956.
13. **Jencks, W. P. and Carriuolo, J.**, *J. Am. Chem. Soc.*, 83, 1743, 1961.
14. **Hestrin, S.**, *J. Biol. Chem.*, 180, 249, 1949.
15. **Capon, B. and Page, T. I.**, *J. Chem. Soc. B*, 741, 1971.
16. **Anderson, E. and Fife, T. H.**, *J. Org. Chem.*, 37, 1993, 1972.
17. **Grunwald, E. and Eustace, D.**, in *Proton Transfer Reactions*, Caldin, E. and Gold, V., Eds., Chapman and Hall, London, 1974.
18. **Bender, M. L. and Komiyama, M.**, *Cyclodextrin Chemistry*, Springer-Verlag, New York, 1978.
19. **Akiyama M., Ohmachi, T., and Ihjima, M.**, *J. Chem. Soc. Perkin Trans. 2*, 1511, 1982.
20. **Skorey, K. I., Somayaji, V., and Brown, R. S.**, *J. Am. Chem. Soc.*, 111, 1445, 1989.

Chapter 2

Stoichiometry and Stability of α-Cyclodextrin Complexes with Aromatic Substrates in Aqueous Solution

Kenneth A. Connors

TABLE OF CONTENTS

I. INTRODUCTION

Cycloamyloses (cyclodextrins) are cyclic oligomers containing six or more D-glucose units linked 1-4. The six-, seven-, and eight-unit substances are called cyclohexaamylose (α-cyclodextrin), cycloheptaamylose (β-cyclodextrin), and cyclooctaamylose (γ-cyclodextrin), respectively. These molecules are doughnut shaped, and their possession of a cavity of fixed size and shape has led to considerable interest in their properties as host molecules for the formation of host-guest inclusion complexes. The introduction of cyclodextrins as enzyme models, by Bender and co-workers in 1967,[1] was an important stimulus to studies of the chemistry of cyclodextrin complexation. From our own viewpoint in pharmaceutics, it seems clear that we must better understand many features of the solution chemistry of cyclodextrin complexes before we can confidently employ them in pharmaceutical dosage forms, an area in which many applications have been proposed. Several years ago, we began systematic studies designed to clarify some issues in the equilibrium solution chemistry of cyclodextrin complexes. These issues include identification of the sites of binding, possible existence of isomeric complexes, occurrence of multiple stoichiometric ratios, estimation of microscopic binding constants, possibility of cooperative binding, and the nature of the binding forces. The present paper brings together our results from many sources to provide a picture of our understanding of these issues.

Throughout this paper, α-cyclodextrin is the *ligand* (L) and the other interactant is the *substrate* (S). Stoichiometric ratios are specified in the order S:L, so SL_2 is a 1:2 complex. Stepwise stability constants are defined as $K_{11} = [SL]/[S][L]$ and $K_{12} = [SL_2]/[S][L]$, where brackets signify molar concentrations; all constants refer to 25°C, ionic strength 0.1 M (unless otherwise stated), in aqueous solution. These constants were measured by spectroscopic, potentiometric, solubility, or competitive spectroscopic techniques.[2] All of the substrates are either disubstituted benzenes or disubstituted biphenyls.

II. THE BINDING-SITE MODEL

Most organic compounds are too large to be completely included in a cyclodextrin cavity, so we focus attention on that portion of a substrate that can enter the cavity; this part of the substrate is called a *binding site*. Disubstituted benzenes and biphenyls possess two potential binding sites. Cyclodextrins as ligands also possess two potential binding sites, namely, the two entrances to the cavity. In a system of a 2-site substrate and a 2-site ligand, there may exist four isomeric 1:1 complexes, four 2:1 complexes, and four 1:2 complexes.[3] For α-cyclodextrin, however, there appears to be little possibility of complexing at the narrow (primary hydroxyl) end of the torus. Thus, α-cyclodextrin behaves as a 1-site ligand, which, with a 2-site substrate, can form only two 1:1 complexes and one 1:2 complex. The equilibria among these species is shown in Scheme I, where XY is a substrate having binding sites X and Y, and ligand binding at a site is shown by a superscript prime.

$$
\begin{array}{ccc}
& X'Y & \\
K_{X'Y} \nearrow & & \nwarrow K_{X'Y'}^{*} \\
XY & & X'Y' \\
K_{XY'} \searrow & & \nearrow K_{X'Y'}^{**} \\
& XY' &
\end{array}
\qquad\qquad \text{Scheme I}
$$

In Scheme I, $K_{X'Y}$ is the microscopic binding constant for the formation of X'Y from XY, and so on. The 1:2 complex X'Y' can be formed by adding a ligand molecule to either of the 1:1 complexes.

For this scheme, Equations 1 and 2 can be obtained,[4]

$$K_{11} = K_{X'Y} + K_{XY'} \tag{1}$$

$$K_{12} = \frac{a_{XY}K_{X'Y}K_{XY'}}{K_{11}} \tag{2}$$

where $a_{XY} = K_{X'Y'}^{*}/K_{XY'} = K_{X'Y'}^{**}/K_{X'Y}$, and K_{11}, K_{12} are the stepwise binding constants defined earlier. The quantity a_{XY} is an interaction parameter that measures the extent of interaction between sites X and Y in 1:2 complex formation. If the sites are independent, $a_{XY} = 1$, but there is in general no restriction on the value of the interaction parameter. We measure the two parameters K_{11} and K_{12}, but our goal is to estimate the three parameters $K_{X'Y}$, $K_{XY'}$, and a_{XY}.

To this formal stoichiometric model we add chemical content with this postulate: complex stability (at a binding site) is enhanced by an increase in electron density at the binding site, it is enhanced by an increase in site polarizability, and it is decreased by an increase in site polarity (in a polar solvent). The first two statements in this postulate assume that induction and dispersion forces contribute to complex stability; the third statement essentially says that the hydrophobic interaction is a more important contributor than is dipole interaction.

As an initial example of the utility of the binding site model, consider the data[3] in Table 1 for substrates having the basic structure $C_6H_5-CH=CH-COX$.

Obviously, the two binding sites are the phenyl ring and the side chain. Molecular models indicate that the 3,5-dimethoxy-substituted ring cannot enter the α-cyclodextrin cavity; hence, 3,5-dimethoxycinnamic acid is a 1-site substrate. As a consequence, one of the microscopic 1:1 binding constants is zero, and the model predicts that $K_{12} = 0$, as observed. Moreover, the observed K_{11} can be assigned to complexing at the side chain. By analogy, in the other substrates, the side-chain binding site is expected to be more strongly complexed than is the phenyl site, with the small K_{12} reflecting this

TABLE 1
Binding Constants for α-Cyclodextrin Complexes of Cinnamate
Substrates (Ionic Strength 0.01 M)

Substrate	X	K_{11}/M^{-1}	K_{12}/M^{-1}	μ/D
Cinnamic acid	OH	2260	60	1.31
3,5-Dimethoxycinnamic acid	OH	1965	0	—
Methyl cinnamate	OCH_3	1200	50	1.95
Benzalacetone	CH_3	105	15	3.34

weak binding at the phenyl group. Of course, the interaction parameter cannot be estimated from these data. The comparatively weak binding of benzalacetone is consistent with the chemical postulate, as shown by the dipole moment data.

III. UNSYMMETRICAL SUBSTRATES

By an unsymmetrical substrate is meant a 2-site substrate whose sites X and Y are chemically different. We consider here substituted benzoic acids, phenols, and anilines. Since these substrates have acid-base character, each substrate can exist in two forms, namely, the conjugate acid and the conjugate base. Thus, the experimental binding constants are written K_{11a}, K_{12a}, K_{11b}, and K_{12b} to distinguish between the acid and base forms of the substrate. If K_a is the acid dissociation constant of the substrate and K_{a11} is the acid dissociation constant of the complexed substrate, then it is easily shown that $K_a K_{11b} = K_{a11} K_{11a}$, i.e., if $K_{11a} > K_{11b}$, then the uncomplexed substrate is a stronger acid than is the complexed substrate (and vice versa). It is a matter of taste whether one regards the relative acidities as a cause and the complex stabilities as the effect, or the other way around. These systems are easily studied by the potentiometric method.[2,4,5]

Table 2 lists binding constants for α-cyclodextrin with *meta*[6]- and *para*[4]- substituted benzoic acids. (Measures of precision of results quoted here can be found in the original papers.) For all of these systems, $K_{12b} = 0$.

Many conclusions can be drawn from Table 2 in conjunction with the binding site model. Some of these are most evident in the form of the Hammett plot of Figure 1. Clearly, $K_{11a} > K_{11b}$ and the Hammett ρ value is negative for K_{11a} but positive for K_{11b}. Making use of the chemical postulates of the model, we conclude that:

1. In the conjugate acid substrates, the principal contributor to K_{11a} is binding at the COOH site, because increased electron withdrawal by X reduces complex stability.

2. In the conjugate base substrates, the only contributor to K_{11b} is binding at the X-site, because $K_{12b} = 0$, and because ionization of the carboxylic acid produces a highly polar site. Thus, K_{11b} is a microscopic binding

TABLE 2
Stability Constants for Substituted Benzoic
Acids, $X-C_6H_4-COOH$

X	K_{11a}/M^{-1}	K_{11b}/M^{-1}	K_{12a}/M^{-1}
	Para Series		
$NHCH_3$	1301	6.1	0
NH_2	1341	9.0	0
OH	1130	16.6	0
OCH_3	884	3.5	0
CH_3	1091	6.6	0
H	722	11.2	0
F	504	14.2	0
CH_3CO	899	60.3	28.8
CN	471	79.2	25.0
NO_2	350	81.0	20.2
	Meta Series		
OH	455	6.3	0
OCH_3	855	14.9	0
CH_3	496	14.9	0
Cl	1165	68.1	0
COOH	233	—	0
CN	365	60.5	0
NO_2	109	29.7	0

constant. The scatter about the K_{11b} Hammett plot exceeds that of the K_{11a} plot because X is, for K_{11b}, both the substituent and the reaction site.

3. Since K_{11a} describes mainly COOH binding, K_{12a} describes mainly X-binding. Thus, K_{12a} is related to K_{11b}, since they both describe X-binding. In the *para* series, there is a rough correspondence; in the *meta* series, steric interference prevents 1:2 complexation.

4. On the average over many complexes, K_{11} *(meta)* $\approx K_{11}$ *(para)* in both the acid and base series.

It is clear that definite conclusions as to preferred binding site and relative complex stability can be reached, but it is also evident that one is, in effect, painting with a broad brush, because of the incursion of geometric factors, because the substituent is not always distinctly different from the reaction site, and because the polarizability and polarity factors are not quantitatively accounted for.

Tables 3 and 4 give data for 4-substituted phenols[7] and 4-substituted anilines.[8] For all of these systems, $K_{12a} = 0$. From Tables 3 and 4, we reach these conclusions:

1. For both K_{11a} and K_{11b}, a trend of increasing complex stability with increasing Hammett σ is seen (interrupted by specific effects such as

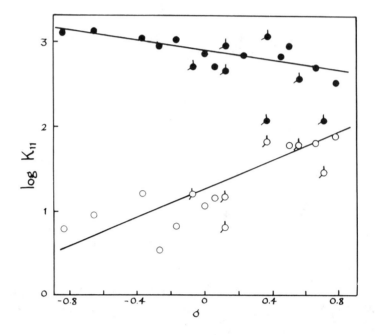

FIGURE 1.   Hammett plots of K_{11a} (filled circles) and K_{11b} (open circles) for 3- and 4-substituted benzoic acids. 3-Substituted substrates are indicated by symbols having dashes oriented at 120°.

TABLE 3
Stability Constants for α-Cyclodextrin Complexes of 4-Substituted Phenols, $X–C_6H_4–OH$

X	K_{11b}/M^{-1}	K_{12b}/M^{-1}	K_{11a}/M^{-1}
OCH_3	2.7	13.3	0
CH_3	13.9	0	0
H	10.9	0	0
COO^-	—	—	16.6
F	15.6	0	0
I	3955	2.4	2316
Cl	487.9	3.1	272
Br	1221	4.7	704
COOH	—	—	1130
CN	662	0	158.3
NO_2	2408	0	245

the high polarizability of halogens.) This implicates the X-site as the major contributor to K_{11a} and K_{11b}.

2.   Since K_{12a} is 0, it follows that K_{11a} is a microscopic binding constant for complexation at the X-site in the conjugate acid substrates. For those systems having $K_{12b} = 0$, K_{11b} is a microscopic constant for binding at X in the base substrates. When $K_{12b} \neq 0$, some binding must occur

TABLE 4
Stability Constants for α-Cyclodextrin Complexes of 4-Substituted Anilines, $X-C_6H_4-NH_2$

X	K_{11b}/M^{-1}	K_{12b}/M^{-1}	K_{11a}/M^{-1}
NH_2	2.3	0	0
OCH_2	6.7	0	0
CH_3	57.6	3.9	37.1
H	8.8	0	0
COO^-	9.0	0	—
Cl	251	0	68.6
COOH	1341	0	—
CN	451	—	—
NO_2	635	—	—

TABLE 5
Comparison of K_{11b} for Anilines and K_{11a} for Phenols

X	K_{11b} (aniline)	K_{11a} (phenol)
OCH_3	6.7	0
CH_3	57.6	0
H	8.8	0
COO^-	9.0	16.6
Cl	251	272
COOH	1341	1130
CN	451	158
NO_2	635	245

(though weakly) at the other binding site. In the phenol series, this only takes place when X can tolerate a positive charge by delocalization, which may counter the electron release by the phenolate site.

3. If these assignments are correct, K_{11a} for a substituted phenol should be similar to K_{11b} for the corresponding aniline. This comparison is shown in Table 5; the agreement is quite reasonable for such a simple argument.

Note that $K_{11b} > K_{11a}$ for phenols, i.e., the ionized form of the substrate "partitions" into the cyclodextrin cavity more extensively than does the neutral form. This behavior has occasioned much discussion with respect to 4-nitrophenol.[7] Table 3 shows that the effect is general. It may be accounted for in this way: as we have seen, binding in phenols occurs, for both the conjugate acid and base forms, at the X-substituent. Because of extensive charge delocalization in phenols, the electron density at site X is greater in the anion than in the neutral form. Moreover, in a neutral phenol (which is a highly dipolar species), there is no binding at the −OH site, whereas some

TABLE 6
Stability Constants and Interaction Parameters for α-Cyclodextrin Complexes with Sym–X–C_6H_4–X

X	K_{11}/M^{-1}	K_{12}/M^{-1}	a_{xx}
NH_2	2.3		
OCH_3	75.4	221	11.7
OC_2H_5	128	326	10.2
I	5060	6250	4.94
Br	913	397	1.74
Cl	232	90	1.55
CO_2H	1344	23.8	0.071
CO_2CH_3	454	106	0.93
$COCH_3$	10.2	—	—
CN	33.1	7.2	0.87
NO_2	35.8	4.6	0.51

binding may occur at the phenolate site, which in the present sense may be less polar than the hydroxy group in the neutral substrate.

IV. SYMMETRICAL SUBSTRATES

A symmetrical substrate is, in the present sense, a 2-site substrate whose two sites are chemically identical. Let such a substrate be denoted XX. Then, for this substrate, Equations 1 and 2 become

$$K_{11} = 2K_{X'X} \tag{3}$$

$$K_{12} = a_{XX}K_{11}/4 \tag{4}$$

For such a system, the microscopic binding constant $K_{X'X}$ and the interaction parameter a_{XX} can be evaluated.

Table 6 gives K_{11}, K_{12}, and a_{XX} values for the complexes of α-cyclodextrin with many sym-1,4-disubstituted benzenes,[9] and Table 7 shows results for sym-4,4′-disubstituted biphenyls.[10]

In the conventional sense of the term, a "reaction series" is a set of compounds having a common reaction site and a variable substituent. In this sense, therefore, these substrates do not constitute reaction series because both the reaction site and the substituent are variable. There is nevertheless a familial relationship among these compounds as a consequence of their size, shape, and aromatic character. It is therefore reasonable to look for patterns in their complexing behavior. One such pattern is found in the dependence of K_{11} on S_0, the molar solubility of the substrate in water. For seven of the benzenes and seven of the biphenyls, Equations 5 and 6 describe these relationships.

TABLE 7
Binding Constants for α-Cyclodextrin Complexes with Sym-4,4′-Disubstituted Biphenyls, $X-C_6H_4-C_6H_4-X$

X	K_{11}/M^{-1}	K_{12}/M^{-1}	a_{XX}
OH	41	345	33
CH_3	1000	123	0.5
H	50	63	5.0
Cl	1030	1620	6.3
Br	4330	5335	4.9
CN	302	59	0.78
NO_2	855	147	0.68
COOH[a]			

[a] $\beta_{12} = K_{11}K_{12} = 2.91 \times 10^7\ M^{-2}$.

$$\log K_{11} = -0.59\log S_0 + 0.40 \text{ (benzenes)} \tag{5}$$

$$\log K_{11} = -0.58\log S_0 - 0.66 \text{ (biphenyls)} \tag{6}$$

The slopes of these lines are not significantly different. These relationships imply the dissolution of a substrate (actually the reverse process, crystallization) is a good model of its inclusion in the cyclodextrin cavity, and the model is equally satisfactory for both series of substrates. The relative positions of these two correlation lines (benzenes lying above the biphenyls) can be accounted for in this way: consider two compounds $X-C_6H_4-X$ and $Y-C_6H_4-C_6H_4-Y$ having the same solubility S_0. We can generally expect that, since their solubilities are the same, group X must be less polar than group Y; hence, we anticipate that the benzene compound will form a stronger complex (in a polar solvent) than will the biphenyl compound. Therefore, the benzene correlation line will lie above the biphenyl line. A second factor is that the portion of the substrate that is not included in the cyclodextrin complex is larger for the biphenyl member of the pair, and (since Y is more polar than X) solvation effects will tend to destabilize the biphenyl complex relative to the benzene complex.

The dicarboxybiphenyl system is unusual in that it was not possible to estimate K_{11}, although the product $K_{11}K_{12}$ could be determined.[10] This means that this system is highly cooperative; the 1:1 complex is not detectable, complexation proceeding essentially completely to the 1:2 stage.

V. MICROSCOPIC BINDING CONSTANTS

For the 1,4-disubstituted benzenes in Table 6, unambiguous microscopic binding constants are available via Equation 3. We earlier identified many systems in Tables 2, 3, and 4 for which K_{11} could be interpreted as essentially a microscopic binding constant. Table 8 gathers these quantities.

TABLE 8

Microscopic Binding Constants $K_{X'Y}$ for 1,4-Disubstituted Benzene-α-Cyclodextrin Systems at 25°C

X	Y	$K_{X'Y}/M^{-1}$	X	Y	$K_{X'Y}/M^{-1}$
NO_2	NO_2	17.9	Cl	NH_3^+	68.6
	COO^-	72.2		Cl	116
	OH	245		OH	272
	NH_2	635		NH_2	251
	O^-	2408		O^-	488
Br	Br	457	I	I	2530
	OH	704		OH	2316
	O^-	1221		O^-	3955
CN	CN	16.6	COOH	COOH	672
	COO^-	71.5		F	504
	OH	158		H	722
	NH_2	451		CH_3	1146
	O^-	662		OCH_3	891
				OH	1115
				NH_2	1341
				$NHCH_3$	1301

Table 8 provides $K_{X'Y}$ data for six true reaction series in which X (the reaction site) is held constant while Y (the substituent) is varied. Hammett plots of these data can be written[9] as:

$$\log K_{X'Y} = \rho_X(\sigma_Y - \sigma_X) + \log K_{X'Y}^0 \qquad (7)$$

where $K_{X'Y}^0 = K_{X'X}$ if the correlation is perfect. (This abscissa scale places the values for all XX-type substrates at $\sigma_Y - \sigma_X = 0$). The slopes, ρ_X, depend upon X, as shown in Equation 8.

$$\rho_X = 0.324 \log K_{X'X} - 1.206 \qquad (8)$$

Equation 8 allows an interesting prediction to be made. According to the postulate that an increase in site electron density favors binding, ρ_X for binding at X cannot be positive, and hence its maximum value is zero; alternatively, the stronger the binding, the less susceptible it is to substituent effects.[9] Placing $\rho_X = 0$ in Equation 8 leads to an estimate of the maximum possible microscopic binding constant at a substituted benzene site. The value is $\log (K_{X'X})_{max} = 3.72$, or $(K_{X'X})_{max} = 5.3 \times 10^3\ M^{-1}$. Since $\rho_X = 0$ under this condition, this is also the maximum value for $K_{X'Y}$.

TABLE 9
Calculation of Isomeric Fractional Composition in XY Systems

X	Y	$K_{X'Y}$	$K_{XY'}$	$f_{X'Y}$	$f_{XY'}$
Cl	Br	116	406	0.22	0.78
Br	I	475	1414	0.25	0.75
CN	COOH	25	431	0.05	0.95
NO_2	COOH	33	281	0.11	0.89
Cl	COOH	88	615	0.13	0.87

We are now in a position to estimate microscopic binding constants for other substrates of the XY type, using Equations 9 and 10,

$$\log K_{X'Y} = \rho_X(\sigma_Y - \sigma_X) + \log K_{X'X} \tag{9}$$

$$\log K_{XY'} = \rho_Y(\sigma_X - \sigma_Y) + \log K_{Y'Y} \tag{10}$$

where $K_{X'X} = K_{11}^{XX}/2$, $K_{Y'Y} = K_{11}^{YY}/2$, and $\rho_X(\rho_Y)$ is calculated with Equation 8. However, this method does not take advantage of data that may be available on XY. An alternative approach is to make use of measurements on the XY system. Define X and Y such that $K_{11}^{YY} > K_{11}^{XX}$. Then calculate $K_{X'Y}$ with Equation 9, and finally, calculate $K_{XY'}$ with Equation 1, using the experimental K_{11}^{XY}.

Now in such a system there may exist the isomeric complexes X'Y and XY'. Since these will have different physical and chemical properties, it is of interest to estimate the solution isomeric fractional composition, which can be expressed:

$$f_{X'Y} = \frac{[X'Y]}{[X'Y] + [XY']} = \frac{K_{X'Y}}{K_{11}^{XY}} \tag{11}$$

$$f_{XY'} = K_{XY'}/K_{11}^{XY} \tag{12}$$

Table 9 shows the results of such calculations. Evidently significant fractions of two isomeric complexes may exist in these systems.

As a further example, consider the cyclodextrin-catalyzed hydrolysis of m-nitrophenyl acetate described by VanEtten et al.[1] and ascribed by them to binding of the nitro site in the cyclodextrin cavity, with consequent positioning of the ester function near the secondary hydroxyls on the rim of the cavity. We use the calculational method given above, with $K_{11}^{XY} = 53 M^{-1}$ as reported[1] and $K_{11}^{XX} = 22 M^{-1}$, where $X = NO_2$ and $Y = OCOCH_3$. The result is $K_{X'Y} = 21 M^{-1}$, $K_{XY'} = 32 M^{-1}$, $f_{X'Y} = 0.4$, and $f_{XY'} = 0.6$. Thus, only about 40% of the 1:1 complex exists as the catalytically productive form, and the actual rate enhancement is more than twice that calculated if isomerism is not taken into account.

VI. THE INTERACTION PARAMETER

Since a_{xx} is a ratio of equilibrium constants ($a_{xx} = K_{x'x'}/K_{x'x}$), it is itself an equilibrium constant; a_{xx} is the equilibrium constant for the disproportionation:

$$2X'X \xrightleftharpoons{a_{xx}} X'X' + XX$$

Thus, if the binding sites are independent, $a_{xx} = 1$, and if $a_{xx} \neq 1$, the sites are not acting independently. Hence, this parameter is a convenient measure with which to investigate the interaction between sites. Several factors may influence the magnitude of the interaction parameter a_{xx}.

The electronic effect of L bound at site X' on the nature of site X — If the sites in XX are electron deficient (as a consequence of electron donation to the ring), upon interaction of one of them with L to give X'X, there will be a partial electron transfer from L to the binding site. This has the effect of increasing the charge density at site X in X'X relative to that at X in XX. Thus, binding of the second ligand will be favored relative to that of the first one, and a_{xx} will be greater than unity. Thus, a_{xx} may be expected to follow a Hammett plot with a negative slope.

The repositioning effect — In the 1:1 complex, the relative position of ligand and binding site is optimal with regard to lowering the total free energy of the system. Formation of the 1:2 complex will result in adjustment of all three molecules to minimize the total free energy, since in the 1:2 complex X'X', the two bound sites are necessarily identical on average. This may require a repositioning of the substrate-ligand orientation that was reached in the 1:1 complex. Any such repositioning must therefore be destabilizing and will therefore lower a_{xx}.

The ligand-ligand interaction effect — In a 1:2 complex, there is a possibility that the facing rims of the two cyclodextrin molecules may interact attractively. Such an effect could only be manifested as 1:2 complex stabilizing (increasing a_{xx}), because any destabilizing repulsive interactions would be accounted for in terms of the repositioning effect.

Hammett plots of log a_{xx} vs. σ_x (for nine members of the benzene series and six of the biphenyls)[10] gave these relationships:

$$\log a_{xx} = -1.29\sigma_x + 0.68 \text{ (benzenes)} \tag{13}$$

$$\log a_{xx} = -1.45\sigma_x + 0.94 \text{ (biphenyls)} \tag{14}$$

Terephthalic acid (X = COOH in the benzene series) falls far below the correlation line of Equation 13. From the above arguments on the factors controlling a_{xx}, it follows that this is a consequence of a major repositioning effect, suggesting that the COOH group is deeply inserted into the cyclodextrin cavity in the 1:1 complex. In the biphenyl series, dicarboxybiphenyl is an outstandingly serious positive deviator (since the 1:1 complex cannot be detected). If, as the terephthalic acid result suggests, the carboxy group is deeply inserted in the 1:1 complex, presumably it occupies a similar position in the dicarboxybiphenyl complex. Addition of a second ligand to terephthalic acid required displacement of the first, giving a low a_{xx} value; but in the dicarboxybiphenyl case, addition of the second ligand must be highly favorable, so much so that the 1:2 complex forms with the virtually complete extinction of the 1:1 complex. It may be inferred that the spacing between sites in the biphenyl complex is optimal for avoiding the repositioning effect and for bringing the ligand-ligand interaction into play.

It is worth noting that it does not necessarily follow that the sites are independent if $a_{xx} = 1$; it is quite possible for the several effects to combine so as to yield this result fortuitously.

REFERENCES

1. **VanEtten, R. L., Sebastian, J. F., Clowes, G. A., and Bender, M. L.,** *J. Am. Chem. Soc.,* 89, 3242, 1967; **VanEtten, R. L., Clowes, G. A., Sebastian, J. F. and Bender, M. L.,** *J. Am. Chem. Soc.,* 89, 3253, 1967.
2. **Connors, K. A.,** *Binding Constants: The Measurement of Molecular Complex Stability,* Wiley-Interscience, New York, 1987.
3. **Rosanske, T. W. and Connors, K. A.,** *J. Pharm. Sci.,* 69, 564, 1980.
4. **Connors, K. A., Lin, S.-F., and Wong, A. B.,** *J. Pharm. Sci.,* 71, 217, 1982.
5. **Connors, K. A. and Lipari, J. M.,** *J. Pharm. Sci.,* 65, 379, 1976.
6. **Pendergast, D. D. and Connors, K. A.,** *Bioorg. Chem.,* 13, 150, 1985.
7. **Lin, S.-F. and Connors, K. A.,** *J. Pharm. Sci.,* 72, 1333, 1983.
8. **Wong, A. B., Lin, S.-F., and Connors, K. A.,** *J. Pharm. Sci.,* 72, 388, 1983.
9. **Connors, K. A. and Pendergast, D. D.,** *J. Am. Chem. Soc.,* 106, 7607, 1984.
10. **Connors, K. A., Paulson, A., and Toledo-Velasquez, D.,** *J. Org. Chem.,* 53, 2023, 1988.

Chapter 3

AN ARTIFICIAL ENZYME

Valerian T. D'Souza

TABLE OF CONTENTS

I. INTRODUCTION

This is a comprehensive discussion of Myron L. Bender's contribution to the understanding of enzymatic catalysis which allowed him to build a successful artificial enzyme.[1] Although some of the interpretations of the results he presented have later been reexamined,[2] the original reasoning for his work has inspired many chemists around the world to continue the work on artificial enzymes.

Enzymes are proteins with catalytic activity that exhibit high specificity and large rate accelerations. In spite of their efficiency, versatility, and economy, there are limitations that make them inadequate for practical use. Their structural complexity makes them difficult to study and their fragility renders them inept for industrial use. Artificial enzymes, constructed from known synthetic precursors, can be used in place of real enzymes to understand their mechanism of action and as catalysts under adverse conditions, depending on their design. Although enzymes are large, complex molecules, their power to catalyze reactions can be attributed mainly to binding and catalysis. Binding is responsible for the specificity of the reaction and brings the substrate in close proximity and in the correct orientation to the active site. The artificial enzymes are designed to mimic the exact nature of the binding subsite in terms of shape, size, and microscopic environment, and the active site in terms of identity of groups, stereochemistry, interatomic distances of various groups, and the mechanism of action of the enzyme.[3] The design and synthesis of a successful artificial enzyme based on chymotrypsin is described here.

II. CHYMOTRYPSIN

Investigations of chymotrypsin[4] have revealed that the binding subsite of chymotrypsin is 10 to 12 Å deep and 3.5 to 4 × 5.5 to 6.5 Å in cross section, which gives a snug fit to an aromatic ring which is 6 Å wide and 3.5 Å thick.[5] It is essentially hydrophobic in nature and is capable of having maximum interaction with an aromatic ring to orient the oxygen atom of Ser-195 for a nucleophilic attack on the carbonyl carbon atom of the ester or amide substrate.

The catalytic site of chymotrypsin has been shown to contain (1) a hydroxyl group (of serine-195), (2) an imidazole group (of histidine-57), and (3) a carboxylate ion (of aspartate-102). Interatomic distances are 2.8 Å between the Ser-195 O^γ and the His-57 $N^{\epsilon 2}$ and 2.65 Å between the His-57 $N^{\delta 1}$ and the Asp-102 $O^{\delta 1}$.

The "proton-transfer relay" system proposed for the mechanism of action of chymotrypsin (Figure 1)[6] consists of two proton transfers, one initiated by carboxylate ion and the other by imidazole. These increase the nucleophilicity of the hydroxyl oxygen atom of serine toward the carbonyl function of the amide or ester substrate bound in the hydrophobic pocket of the enzyme to

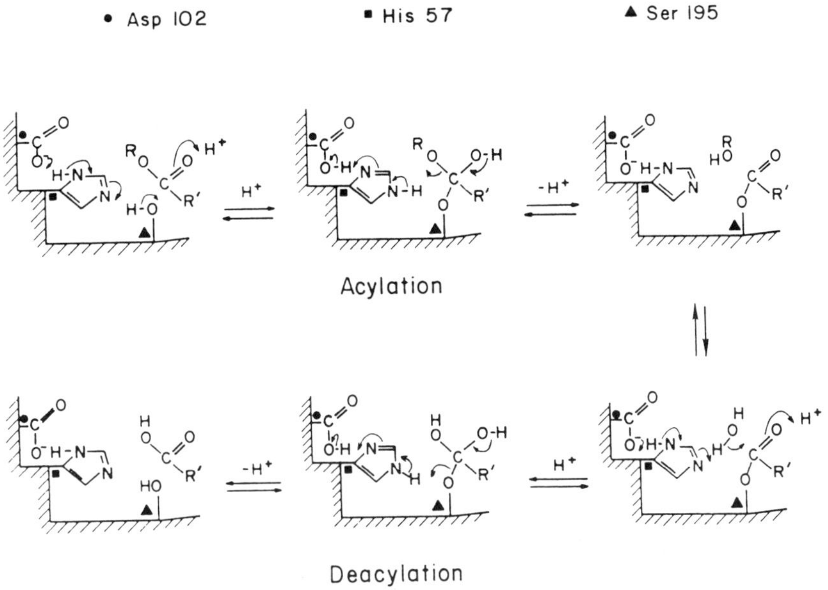

• Asp 102 ■ His 57 ▲ Ser 195

Acylation

Deacylation

FIGURE 1. Mechanism of action of chymotrypsin.

give an acyl-enzyme intermediate. Similarly, deacylation occurs via two proton transfers, increasing the nucleophilicity of the hydroxyl group of water, which attacks the carbonyl group of the acyl-enzyme ester. This mechanism is controversial,[7] and the participation of carboxylate ion has been questioned.[8] However, the experiments with acyl-enzyme models lend credence to the participation of carboxylate ion.

III. DESIGN OF THE ARTIFICIAL ENZYME

The artificial chymotrypsin was designed to contain a hydrophobic pocket to act as a binding subsite, attached to a hydroxyl group, an imidazole group, and a carboxylate ion (the catalytic subsite) placed at right distances and correct stereochemistry to participate in a "proton-transfer relay system".

A. BINDING

The molecule on which to base a model of chymotrypsin should have a cavity that provides maximum hydrophobic interaction with a substrate to form complexes, fits the aromatic ring of the substrate, and orients the carbonyl of the bound substrate toward an oxygen atom (of serine in the real enzyme) for nucleophilic attack.

Cyclodextrins consisting of six (in α), seven (in β), or eight (in γ) glucose units linked together by α-1,4-glycosidic linkages in a cyclic fashion to form

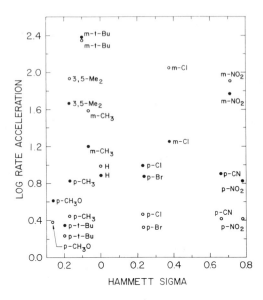

FIGURE 2. Hammett plot for the acceleration of the rate of hydrolysis of substituted phenyl esters by cyclodextrins. (Reprinted with permission from D'Souza, V. T. and Bender, M. L., *Acc. Chem. Res.*, 20, 146, 1987. Copyright 1987, American Chemical Society.)

torus-like structures turned out to be the molecules of choice of this venture.[9] All their secondary hydroxyl groups at the 2- and 3-positions of glucose units are arranged in the more open end and primary hydroxyl groups at the 6-position are located at the other end. The interior of the cavity, consisting of a ring of C–H groups, a ring of glycosidic oxygen atoms, and another ring of C–H groups, is hydrophobic in nature. The inner diameters of the cavities are approximately 4.5 Å in α-cyclodextrin, 7.0 Å in β-cyclodextrin, and 8.5 Å in γ-cyclodextrin; α- and β-cyclodextrins give a snug fit for an aromatic ring.

Aromatic rings were shown to bind to cyclodextrins with dissociation constants varying from 10^{-2} to 10^{-3} M, depending on the substituents on the ring. Hydrophobic substituents on the phenyl ring led to a tighter binding, indicating that the binding was essentially due to hydrophobic interactions. Investigations on the effect of cyclodextrin on the hydrolysis of *p*-nitrophenyl acetate showed that, similar to enzymes, it followed saturation kinetics data which were treated by a variant of Michaelis-Menten kinetics, indicating that the accelerations in the presence of cyclodextrin were due to the formation of Michaelis-Menten-type complexes. The investigations of the stereochemistry of binding by the effect of cyclodextrins on hydrolysis of a series of substituted phenyl acetates produced "the world's worst Hammett plot", compared to the normal Hammett plot produced by the effect of the monomeric methylglucoside (Figure 2). The hydrolytic acceleration by cyclodextrin on *meta*-substituted phenyl esters is larger than that on the corresponding *para*-substituted phenyl esters, indicating that the electronic effects are not important. This in conjunction with the models of binding of *m-tert*-butylphenyl acetate and *p-tert*-butylphenyl acetate by the cyclodextrin (Figure 3) emphasizes the fact that the stereochemistry of binding by cyclodextrin is important, as Emil Fisher predicted in his "lock and key" theory of enzymatic action in 1894. As shown by the models, the binding of *meta*-substituted phenyl acetates to cyclodextrin orients the carbonyl carbon atom of the ester substrate toward the oxygen atoms of the secondary hydroxyl groups of the cyclodextrin for nucleophilic attack, whereas the complexes of *para*-substituted phenyl esters orient the carbonyl carbon atom far from the secondary or the primary hydroxyl groups. Cyclodextrins mimic the binding in chymotrypsin because they have hydrophobic nature and show orientational effects.

B. CATALYSIS

The common feature in both the acylation and deacylation steps of the chymotrypsin mechanism is that the negative charge from the carboxylate ion is transferred to the oxygen atom of the carbonyl function and concurrently the proton is transferred from the hydroxyl group of serine in acylation, or of water in deacylation, via the imidazole group to the carboxylate ion during the formation of the tetrahedral intermediate.

Models of the acyl enzyme were first synthesized to test these mechanistic features, excluding the contributions from the binding part of the enzymatic catalysis. The model of the acyl-enzyme consisted of a norborane backbone with an imidazole group in the *endo* 5-position and a cinnamoyl ester group in the *endo* 2-position.[10] The positions were chosen so that the imidazole group would not act as a nucleophilic catalyst, but allow a water molecule between the imidazole group and the carbonyl carbon atom of the ester function and thus act as a general base catalyst. The deacylation of this model of acyl-enzyme ester showed, by the D_2O effect, that the imidazole group acts as general basic catalyst and not a nucleophilic catalyst.[11] Computer-aided structures of this model showed that the distances between the carbonyl carbon atom of the model of acyl-enzyme (ester function) and the two nitrogen atoms of the imidazole group were 2.43 and 2.80 Å. However, if a water molecule is inserted between the imidazole group and the ester group, the shortest distance between the nitrogen atom of the imidazole group and the hydrogen atom of the water molecule is 1.60 Å, and the distance between the carbonyl carbon atom of the ester group and the oxygen atom of the water molecule is 1.37 Å. These distances support the experimental evidence that the imidazole acts as a general basic catalyst. The most interesting feature of this model is that the rate of deacylation increased tremendously in the presence of benzoate ion. It was also observed that the rate of deacylation in the presence of benzoate ion is further increased by the addition of dioxane, which was used to simulate the apolar nature of the active site of chymotrypsin. The ratio of the rates in the presence of 0.5 *M* benzoate ion to that in the absence

FIGURE 3.   CPK models of cyclodextrin complexes with (A) *p-tert*-butylphenyl acetate and (B) *m-tert*-butylphenyl acetate. (Reprinted with permission from D'Souza, V. T. and Bender, M. L., *Acc. Chem. Res.*, 20, 146, 1987. Copyright 1987, American Chemical Society.)

of benzoate ion resulted in a 2500-fold acceleration at a dioxane mole fraction of 0.42.[12] Such an increase in the rate of deacylation in the presence of the apolar environment and the third component, carboxylate ion, of the active site of chymotrypsin indicated that this is a good model for the charge relay system.

However, this carboxylate ion in this model is an intermolecular catalyst, whereas chymotrypsin employs an intramolecular carboxylate ion. The advantage

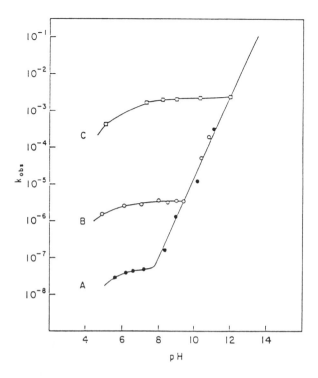

FIGURE 4. Hydrolysis of acyl-enzyme models. (A) Without the carboxylate ion; (B) with intermolecular carboxylate ion; (C) with intramolecular carboxylate ion. (Reprinted with permission from D'Souza, V. T. and Bender, M. L., *Acc. Chem. Res.*, 20, 146, 1987. Copyright 1987, American Chemical Society.)

of intramolecular catalysis over intermolecular catalysis has been well established. Thus, it was imperative to build a model of chymotrypsin with an intramolecular carboxylate ion. This was achieved by synthesizing *endo,endo*-5-[2-(2-carboxyphenyl)-4(5)-imidazolyl]bicyclo[2.2.1]hept-2-yl *trans*-cinnamate. This model has a rate of hydrolysis 154,000 faster than norbornyl cinnamate ester and is only 18-fold slower than deacylation of real *trans*-cinnamoylchymotrypsin (Figure 4). A mechanism similar to that of deacylation of acyl chymotrypsin can be proposed to explain this acceleration (Figure 5). It was suggested that if the same differential solvating system as exists in real chymotrypsin can be mimicked in the active site of the artificial acyl enzyme, then one could expect the artificial acyl-chymotrypsin to deacylate at the same rate as the real acyl chymotrypsin.[13] Thus, a model of the catalytic subsite of chymotrypsin to which a binding subsite could be attached was then synthesized.

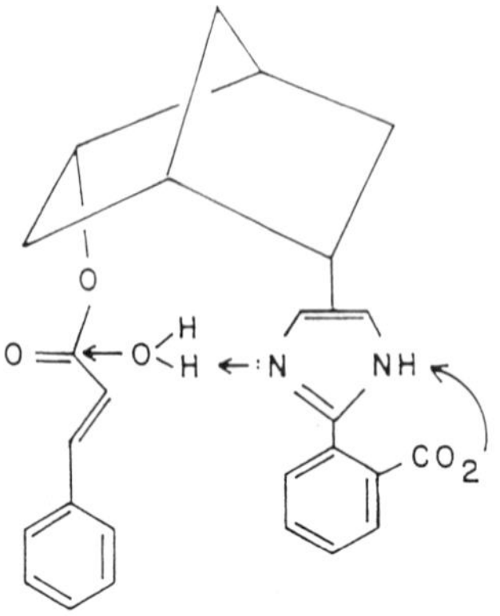

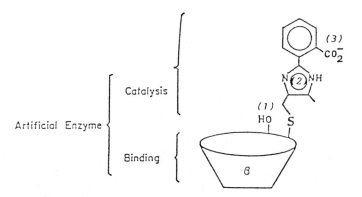

FIGURE 5. Mechanism of action of the model of acyl chymotrypsin. (Reprinted with permission from D'Souza, V. T. and Bender, M. L., *Acc. Chem. Res.*, 20, 146, 1987. Copyright 1987, American Chemical Society.)

FIGURE 6. Design of the artificial chymotrypsin. (Reprinted with permission from D'Souza, V. T. and Bender, M. L., *Acc. Chem. Res.*, 20, 146, 1987. Copyright 1987, American Chemical Society.)

IV. ARTIFICIAL CHYMOTRYPSIN

The combination of the two essential features of enzyme action binding subsite (cyclodextrin) and the catalytic subsite (*o*-imidazolylbenzoic acid) of chymotrypsin should produce an artificial chymotrypsin (Figure 6). Cramer had demonstrated that inserting an imidazole group on the primary side of cyclodextrin led to a fair mimic of chymotrypsin.[14] The artificial chymotrypsin was designed to have the catalytic subsite on the secondary side of cyclodextrin since it was established that the bound substrate would have its carbonyl function at this side of cyclodextrin.

TABLE 1
Hydrolysis of Esters by Chymotrypsins

Enzyme	Substrate	pH	$k_{cat} \times 10^{-2}$, s^{-1}	$K_m \times 10^{-5}$, M	k_{cat}/K_m $M^{-1} s^{-1}$
Chymotrypsin	p-Nitrophenyl acetate	8.0	1.1	4.0	275
Artificial chymotrypsin	m-$tert$-Butylphenyl acetate	10.7	2.8	13.3	210

The synthesis was achieved by the reaction of β-cyclodextrin-2,3-epoxide (obtained from the reaction of β-cyclodextrin 2-tosylate by the reaction of ammonium bicarbonate) with o-[4(5)-mercaptomethyl-4(5)-methylimidazol-2-yl]benzoic acid (obtained by hydrolysis of the corresponding thioacetate in ammonium bicarbonate).

These enzymes models were examined for their catalytic activity in ester hydrolysis. Since m-$tert$-butylphenyl acetate is known to bind well to the β-cyclodextrin cavity, it was decided to examine the rate of hydrolysis of this ester in the presence of β-artificial enzyme. The hydrolysis of more than 10 mol of substrate by 1 mol of artificial enzyme indicates that there is turnover. The results given in Table 1 for β-artificial enzyme obtained from Lineweaver-Burk plots indicate that both the artificial and real enzymes are comparable in their catalytic activity both in rate and in binding constants. The second-order constants (k_{cat}/K_m), the most important enzymatic rate constants, indicate that the artificial enzyme is as efficient as the real enzyme in its catalytic activity. The solvent isotope effect ($k_{H_2O}/k_{D_2O} = 3$) shows that the hydrolysis of m-$tert$-butylphenyl acetate is catalyzed by a general basic mechanism rather than a nucleophilic attack by imidazole. Similar arguments have been put forward to explain the general base character of real chymotrypsin catalysis. Interestingly, the three artificial enzymes have different specificities. Whereas α- and β-artificial enzymes are better than γ-artificial enzyme in phenyl ester hydrolysis, γ-artificial enzyme hydrolyzes tryptophan ethyl ester faster than the α- and β-artificial enzymes. In the hydrolysis of p-$tert$-butylphenyl trimethylacetate by the artificial enzyme, a two-phase reaction similar to the hydrolysis of p-nitrophenyl trimethylacetate by real chymotrypsin[15] was observed (Figure 7).[16]

This is an indication that the artificial enzyme, like real chymotrypsin, involves a pre-steady-state acylation (the curved portion), and a steady-state deacylation and turnover (the straight portion). From the results obtained so far, the mechanism of action of the miniature organic model of chymotrypsin can be compared to the mechanism of action of real chymotrypsin (Figure 8). The mechanism assumes that 1:1 complexes seen in simpler reactions of cyclodextrin are still operative.

There are differences between this miniature organic model of chymotrypsin and real chymotrypsin, both theoretically as well as in practice. The pH maximum for chymotrypsin is 7.9, dependent on the ionization of imidazole

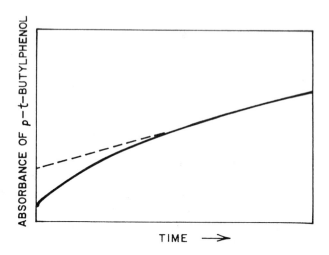

FIGURE 7. Artificial chymotrypsin-catalyzed hydrolysis of *m-tert*-butylphenyl trimethylacetate showing acylation and deacylation. (Reprinted with permission from D'Souza, V. T. and Bender, M. L., *Acc. Chem. Res.*, 20, 146, 1987. Copyright 1987, American Chemical Society.)

and on the conformation determined by a lysine ionization, whereas the pH maximum for the model is beyond 10, determined by both the ionization of imidazole and the ionization of the secondary hydroxyl groups of the cyclodextrin. One of the most important differences between the real and artificial enzymes is their stability. Chymotrypsin can undergo two kinds of inactivation: reversible inactivation and irreversible inactivation. Chymotrypsin, like other enzymes, depends on its conformation for its catalytic activity, which is dictated by hydrogen bonding, van der Waals interactions, configurational entropy, hindered rotation, permanent dipole interactions, electrostatic effects, electronic situations, and interaction of the protein with water. Forces such as temperature that disturb these interactions tend to change the conformation of the enzyme and thus inactivate the enzyme. However, the conformation of the artificial enzyme (the cyclodextrin part) is not affected by such forces and is thus active at elevated temperatures.

Figure 9 shows that real chymotrypsin has a temperature maxima around 45°C, and after 55°C the protein begins to precipitate and is rendered inactive, whereas the activity of the artificial enzyme keeps increasing to at least 80°C.

Irreversible inactivation of chymotrypsin is brought about by disruption of the protein by cleavage of peptide linkages or by blockage of the three groups of the catalytic subsite. Since chymotrypsin consists of 245 amino acids, several of which (tyrosine, tryptophan) are natural substrates for chymotrypsin, it undergoes self-proteolysis (cannibalistic denaturation) at its active pH range. The peptide linkages are also cleaved by hydroxide ion at high pH range to inactivate the enzyme (Figure 10).

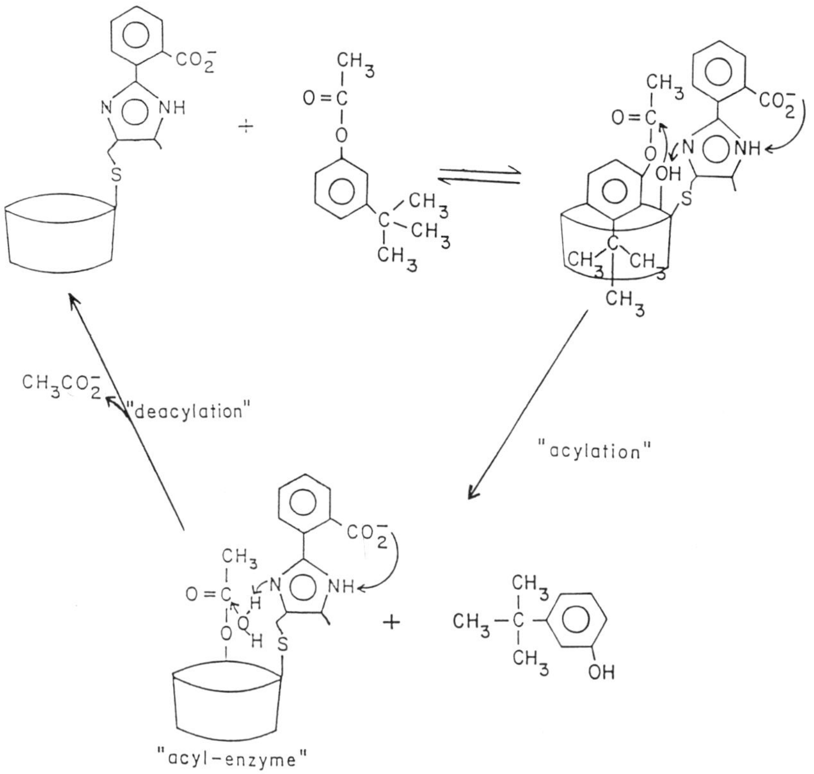

FIGURE 8. Mechanism of action of the artificial chymotrypsin.

However, the artificial enzyme is made up of glucose units with α-1,4-glycosidic linkages. These are very stable under high pH conditions and do not undergo cannibalistic denaturation (since the artificial enzyme is not a substrate for itself). Only at very low pH can glycosidic bonds be broken to inactivate the artificial enzyme (Figure 10). However, the conditions (acidity and temperature) required for such cleavage are too strenuous to hinder practical use of the artificial enzyme. Thus, one of the greatest limitations of natural enzymes, i.e., instability, can be overcome by artificial enzymes.

Despite these differences, chymotrypsin (mol wt 24,800) can be mimicked in an abbreviated form (mol wt 1365), in terms of its two essential features, binding and catalysis.

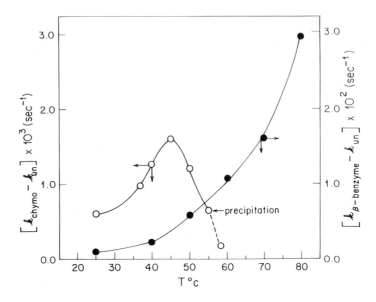

FIGURE 9.   Thermal stabilities of real and artificial chymotrypsins. (Reprinted with permission from D'Souza, V. T. and Bender, M. L., *Acc. Chem. Res.*, 20, 146, 1987. Copyright 1987, American Chemical Society.)

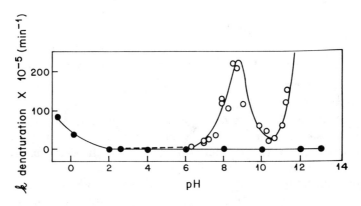

FIGURE 10.   pH stabilities of real and artificial chymotrypsins. (Reprinted with permission from D'Souza, V. T. and Bender, M. L., *Acc. Chem. Res.*, 20, 146, 1987. Copyright 1987, American Chemical Society.)

REFERENCES

1. D'Souza, V. T., Hanabusa, K., O'Leary, T., and Bender, M. L., *Biochem. Biophys. Res. Commun.*, 129, 727, 1985.
2. Breslow, R. and Chung, S., *Tett. Lett.*, 30, 4357, 1989.
3. D'Souza, V. T. and Bender, M. L., *Acc. Chem. Res.*, 20, 146, 1987.
4. Bender, M. L. and Kezdy, F. J., *Annu. Rev. Biochem.*, 34, 49, 1965.
5. Steitz, T. A., Henderson, R., and Blow, D. M., *J. Mol. Biol.*, 46, 337, 1969.
6. Walsh, C., *Enzymatic Reaction Mechanisms*, W. H. Freeman, San Francisco, 1979.
7. Roberts, J. D. and Kanamori, K., *Proc. Natl. Acad. Sci. U.S.A.*, 77, 3095, 1980.
8. Hamilton, S. E. and Zerner, B. J., *Am. Chem. Soc.*, 103, 1827, 1981.
9. Bender, M. L. and Komiyama, M., *Cyclodextrin Chemistry*, Springer-Verlag, New York, 1978.
10. Utaka, M., Takeda, A., and Bender, M. L., *J. Org. Chem.*, 39, 3772, 1974.
11. Komiyama, M., Roesel, T. R., and Bender, M. L., *Proc. Natl. Acad. Sci. U.S.A.*, 74, 2634, 1977.
12. Komiyama, M., Bender, M. L., Utaka, M., and Takeda, A., *Proc. Natl. Acad. Sci. U.S.A.*, 74, 2634, 1977.
13. Mallick, I. M., D'Souza, V. T., Yamaguchi, M., Lee, J., Chalabi, P., Gadwood, R. C., and Bender, M. L., *J. Am. Chem. Soc.*, 106, 7252, 1984.
14. Cramer, F. and Mackensen, G., *Angew. Chem. Int. Ed. Engl.*, 5, 601, 1966.
15. Bender, M. L., Kezdy, F. J., and Wedler, F. C., *J. Chem. Ed.*, 44, 84, 1967.
16. Bender, M. L., *J. Inclusion Phenom.*, 2, 433, 1984.

Chapter 4

MECHANISMS OF CATALYSIS AND INHIBITION OF β-LACTAMASE

Anthony L. Fink

TABLE OF CONTENTS

I. INTRODUCTION

β-Lactamases hydrolyze the amide bond of β-lactam antibiotics, thereby rendering them ineffective. They play a major role in the resistance of pathogenic organisms to β-lactam antibiotics, especially in hospital infections. β-Lactamases are very efficient enzymes, with turnover numbers of as much as several thousand per second for some substrates. Until recently, investigations of the catalytic mechanism have been hindered by the lack of a high-resolution crystallographic structure. The structure of the enzyme from *Staphylococcus aureus* has now been reported.[26] The major role which these enzymes play in resistance to β-lactam antibiotics makes an understanding of their mechanism of action important for the design of mechanism-based antibiotics.

From the crystal structure, the likely essential catalytic residues are Ser-70, Lys-73, Lys-234, and Glu-166. With the exception of the serine, the role of the other active-site residues is unknown, and relatively little is understood regarding the details of the catalytic mechanisms of these enzymes. There has been considerable uncertainty over the years as to whether class A β-lactamase catalysis occurred via an acyl-enzyme intermediate, or whether the mechanism was general acid/base catalyzed without a covalent enzyme-substrate intermediate. The observed absence of transfer reactions in the presence of nucleophiles such as methanol or hydroxylamine is consistent with the latter mechanism. The covalent involvement of serine has been implicated in the interactions with a number of inhibitors.[7,13,14,23,32] Site-directed mutagenesis studies have also shown that Ser-70 is an essential catalytic residue,[16] although the thiol analog has some catalytic activity.[36,37] In addition, several investigations of Class A β-lactamases suggest that the catalytic reaction involves the intermediacy of a covalent intermediate.[1,2,9,23] The catalytic mechanism may thus be represented by Equation 1.

$$\text{E} + \text{S} \underset{}{\overset{K_s}{\rightleftharpoons}} \text{ES} \underset{k_{-2}}{\overset{k_2}{\rightleftharpoons}} \text{EA} \overset{k_3}{\longrightarrow} \text{E} + \text{P} \tag{1}$$

The short lifetime of enzyme-substrate complexes under normal conditions renders their study difficult. By using subzero temperatures to slow the catalytic reaction, it is possible that intermediates may be accumulated sufficiently to permit some structural information to be obtained.[8,20] Various cryosolvents were investigated for their suitability in cryoenzymological experiments with Class A β-lactamases from *S. aureus, Bacillus cereus,* and *B. licheniformis.* Based on the minimal effects on the catalytic and structural properties of the enzyme, ternary solvents containing ethylene glycol, methanol, and water were found most suitable.[10,40] The interaction of β-lactamase with a number of substrates was studied at subzero temperatures. In general, the reaction profiles were similar to those in aqueous solution at above-zero temperatures, with the exception of the slower rates. For cephalosporin substrates, such as (2[{p-dimethylaminophenyl}azo]pyridinio)cephalosporin (PADAC), in which the 3'-substituent may leave to form a more stable form of the acyl enzyme, this intermediate could be readily stabilized at subzero temperatures.

In the work to be discussed here, we show that an acyl-enzyme intermediate involving Ser-70 is formed, that Lys-234 is important for substrate binding, and that the mechanism of inhibition by penicillin sulfones involves both chemical and physical (conformational) steps.

II. TRAPPING THE ACYL ENZYME AT SUBZERO TEMPERATURES

The values of k_{cat} and K_m for Class A β-lactamase-catalyzed hydrolysis of furylacryloylpenicillin in aqueous solution are similar to those for benzyl-penicillin, indicating that it is a good substrate for these enzymes. At $-38°C$ in 20% ethylene glycol/50% methanol, pH* 5.0, we obtained values of k_{cat} = 0.08 s^{-1} and K_m = 90 μM with the staphylococcal β-lactamase. This substrate was well-behaved in reactions carried out at subzero temperatures, had a substantial absorbance change associated with the reaction, and had a reasonably low K_m, allowing for substrate saturation at subzero temperatures, and hence was a good candidate for attempts to trap the putative acyl enzyme at subzero temperature.

The strategy for these experiments was as follows. If the deacylation step were rate limiting (or even partially rate limiting), then with saturating substrate concentrations the majority of the enzyme would be in the form of the acyl enzyme under the low-temperature steady-state conditions. At suitably low temperature, the rate of breakdown of the intermediate would be quite slow. Previous investigations have shown that β-lactamase adopts a nonnative, noncatalytically active conformation at pH values below 3,[5,24] and that this is also true in the presence of methanol.[41] Depending on the exact conditions (pH, temperature, ionic strength, and cosolvent concentration), this process can be quite fast even at subzero temperatures. Thus, if the pH was rapidly dropped after the steady state was established at the low temperature, the acyl-enzyme intermediate might convert into the nonactive conformation at a rate faster than the slow rate of deacylation. Previous studies on the putative acyl-enzyme intermediate have shown that it was relatively stable at low pH.[9]

Various means of dropping the pH were investigated. The one finally chosen involved the addition of 0.5 M trifluoroacetic acid in cryosolvent to bring the pH* to 2. In a typical trapping experiment, furylacryloylpenicillin (30 mM) was mixed with β-lactamase (10 μM) in 20% ethylene glycol/50% methanol, pH* 5.0 at $-40°C$. After 5 s, the reaction was quenched by the addition of TFA to bring the pH* to 2. After removal of the majority of the substrate and product by centrifugal minigel ion-exchange chromatography, the eluant was analyzed by HPLC. As shown in Figure 1A, the furylacryloyl chromophore coeluted with the enzyme. Since the chromatographic conditions are such that

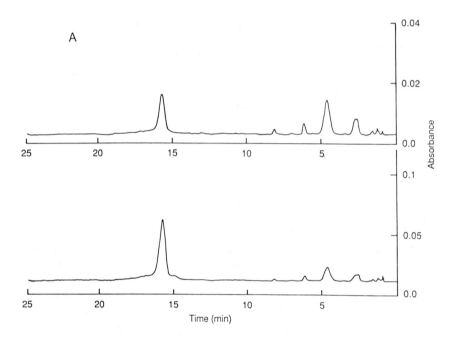

FIGURE 1.   (A) HPLC analysis (C3 reverse-phase column, acetonitrile gradient) of the eluant of a minigel (DEAE Sephadex)[21] separating the products of the reaction of *B. cereus* β-lactamase I with FAP at −40°C at pH* 5.5 in 50% methanol-20% ethylene glycol cryosolvent. The reaction was quenched with 0.5 *M* TFA after 10 s, bringing the pH* to 2. The upper trace monitors the furylacryloyl chromophore at 305 nm, the lower the enzyme at 280 nm. The peak at 3 min corresponds to the product, furylacryloylpenicillanoic acid, the peak at 4.5 min corresponds to unreacted substrate, and the peak at 16 min corresponds to the enzyme. The enzyme fraction was collected, treated with pepsin, and rechromatographed. The labeled peptide from HPLC was collected, dried down, and treated with trypsin. (B) HPLC analysis of the labeled tryptic peptide following pepsin and trypsin digestion of the trapped acyl enzyme. The labeled peptide (9 min) was subjected to amino acid analysis using *o*-phthalaldehyde. The upper trace was monitored at 305 nm to detect the furylacryloyl group, the lower at 215 nm to detect peptides. (Reprinted with permission from Virden, R., Tan, A. K., and Fink, A. L., *Biochemistry, 29,* 145, 1990. Copyright 1990, American Chemical Society.)

the protein should be unfolded,[24] this indicates a covalent bond between the furylacryloylpenicillin and the enzyme. Based on the extinction coefficient of the furylacryloyl chromophore, we calculate a 0.8:1 stoichiometry between FAP and the enzyme.

Slow loss of the chromophore was noted on storage of the acyl enzyme at pH 2, 4°C; e.g., after 18 h, about 25% of the label had been released. If the acyl-enzyme sample was adjusted to pH 5.8 and incubated at 18°C for 30 min, and then reanalyzed by HPLC, it was found that all the label had been released from the acyl enzyme. If this experiment was repeated in the presence of 2 *M* urea, to maintain the enzyme in a noncatalytically active conformation,

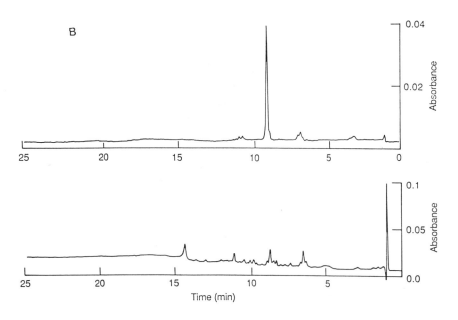

FIGURE 1 (continued).

50% of the label was retained. Thus, restoring the acyl enzyme to native-like conditions resulted in deacylation of the acyl enzyme, whereas maintaining it under nonnative conditions resulted in slow loss of the label, presumably due to acid- or base-catalyzed hydrolysis of the exposed ester linkage in the acyl enzyme. Inactivation of the enzyme by clavulanic acid prior to incubation with the substrate at −40°C led to no incorporation of the label, as determined by the HPLC analysis of the minigel eluant. Similarly, if the substrate was added after the pH was adjusted to 2 at −40°C, no label was incorporated in the enzyme.

The acyl-enzyme trapping experiments reveal a number of interesting points. At the subzero temperatures, the rate of acid denaturation must exceed the rate of deacylation, resulting in the quenching of the acyl-enzyme intermediate. The close to 1:1 stoichiometry indicates not only that the catalytic mechanism involves the covalent acyl enzyme, but also that the deacylation rate is either entirely or predominantly rate limiting. Similar data were obtained with the enzyme from *B. cereus*.

III. DETERMINATION OF THE COVALENTLY MODIFIED RESIDUE IN THE ACYL ENZYME

A sample of the acyl enzyme from the reaction with FAP was prepared as outlined above. The labeled fraction from the HPLC experiment was dried down, taken up in pH 2 buffer, and digested with pepsin. Samples were analyzed by HPLC at various times during the digestion. Three labeled peaks were detected at 305 nm. An initial major peak at RT = 17.41 min was gradually replaced by two with RT = 14.17 and 14.71 min. Over longer time periods, the peak at 14.71 min converted into the one at 14.17 min. Amino acid analysis of the fractions corresponding to the latter two peaks indicated that they contained a large peptide consistent with the sequence from residues 41 to 89 of β-lactamase. These fractions were collected from the HPLC and the pH adjusted to 7.8. The samples were then digested with trypsin and reanalyzed by HPLC. A single labeled peptide was obtained (Figure 1B). Both fractions from the pepsin digestion gave the same labeled peptide on trypsin digestion, with RT = 8.46 min. Amino acid analysis of this peptide indicated that it corresponded to the sequence from Phe-66 to Lys-73. Based on previous reports that various inhibitors modify Ser-70, and that conversion, by site-directed mutagenesis, to other residues except Cys leads to lack of activity, we therefore conclude that the covalent linkage between furylacryloylpenicillin and the enzymes involves Ser-70.

IV. TRAPPING THE ACYL ENZYME AT 25°C

Although FAP is one of the best substrates for the staphylococcal β-lactamase, its turnover number is not very large (k_{cat} = 118 s^{-1} at 25°C, pH 6.5). We thought it possible, therefore, that the competition between acid quenching and deacylation for the acyl enzyme might favor the denaturation even at 25°C. Consequently, we repeated the above acyl-enzyme trapping experiment at 25°C rather than at −40°C. After the acid quench, the sample was analyzed by HPLC and showed the covalent attachment of the FAP. The stoichiometry was calculated to be 0.8:1. Control experiments in which the enzyme was first inactivated by clavulanic acid showed no evidence for covalently bound FAP. Confirmation of the presence of a covalent link was obtained in experiments in which the initially isolated putative acyl enzyme was denatured with 2 *M* urea prior to HPLC analysis, which still showed the presence of the FAP attached to the enzyme. Digestion of the acyl enzyme with pepsin and trypsin yielded the same labeled peptide as from the subzero temperature experiment, confirming the same acyl-enzyme intermediate.

The fact that the acyl enzyme can be trapped in this manner at 25°C means not only that deacylation must be at least partially rate limiting, but also that the partitioning between deacylation and acid-quenched denaturation must favor the latter. This is in contrast to the case with the enzyme from *B. cereus* (k_{cat} > 2000 s^{-1}) in which the acyl enzyme can be trapped at subzero temperature but not at 25°C.[10]

V. ROLE OF LYS-234

From studies of conserved residues of related enzymes, chemical modification experiments, and recent structural data, the following groups are conserved in the active-site region of Class A β-lactamases and are believed to be important for catalysis: (1) Ser-70, which is thought to act as a nucleophile in attacking the β-lactam bond to form a covalent penicilloyl-enzyme intermediate (the acyl enzyme), (2) Lys-73 and Glu-166, which may act to facilitate the transfer of protons in the acylation and deacylation steps, and (3) Lys-234, which has been postulated to serve as an electrostatic anchor for the substrate carboxylate.[26] It is clear that the catalytic mechanisms of the cell-wall D-Ala-D-Ala peptidases must also be similar, based on the high degree of sequence and three-dimensional homology with the β-lactamases.[29,30]

In order to determine whether the postulated role of Lys-234 was correct, we used site-directed mutagenesis to convert it into a glutamate residue. If Lys-234 were important in substrate binding, this change should lead to major effects. Similarly, if its role were as an acid or base catalyst, this change should lead to substantial perturbations of the catalytic mechanism. These experiments were done with the enzyme from *B. licheniformis* cloned and expressed in *B. subtilis*.

The high-resolution structure of β-lactamase shows that Lys-234 is mostly buried below the floor of the active-site depression, with the ammonium group penetrating the floor of the active-site cavity at the closed end. Model building studies based on the requirements of electrostatic, hydrophobic, and shape complementarity suggested that the only good candidate for electrostatic interaction with the substrate carboxylate was Lys-234.[26] Consequently, the postulated role of this conserved residue is to facilitate substrate binding and orientation by formation of a salt bridge with the substrate carboxyl group. It has long been known that a specificity requirement for β-lactamase substrates is a negative charge in the vicinity of C3, although the lactone of deacetylcephalosporin C is a substrate for β-lactamase I of *B. cereus*,[33] with k_{cat} 30% that of benzylpenicillin and $K_m = 2.0$ mM. Interestingly, the non-lactone form is a very poor substrate, with $k_{cat} < 0.1\%$ that of benzylpenicillin. It is possible that these effects are connected with the open or closed conformation of the cephem ring.[11] In the case of the related *Streptomyces* R61 penicillin-sensitive DD-peptidase, comparison of the interaction of deacetylcephalosporin C and the lactone has revealed that an enzyme group of pK $\cong$ 9 may form an ion pair with the substrate carboxylate in substrate binding.[39]

The characterization of the physical properties of the mutant demonstrate that the conversion of Lys-234 into Glu-234 causes negligible effect on the structural properties of the protein. Far-UV circular dichroism was used to compare the secondary structures of the two proteins. The spectra were superimposable, indicating no differences in secondary structure. Similarly, the

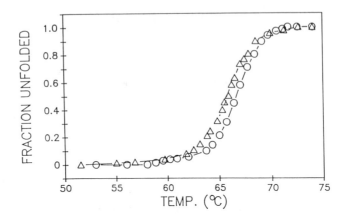

FIGURE 2.   Thermal denaturation curves for wild type (△) and K234E mutant (○). The reaction was followed by tryptophan fluorescence emission at 335 nm at pH 6.0. The enzyme concentrations were 3.2 μm.

tryptophan fluorescence emission spectra were identical for the two proteins, indicating similar tertiary structures. As a further corroboration the susceptibility to proteolysis was investigated. With thermolysin, an identical rate $(4.7 \times 10^{-4} \text{ s}^{-1}$ at 55°C) of loss of catalytic activity was found for both wild type and mutant. Another indication that the mutation has negligible effect on the structure of the protein is found in the thermal stability curves. The values for the T_m were 65.7 and 66.5°C for wild type and mutant, respectively, at pH 6.0 (Figure 2). The isoelectric points were determined using analytical IEF; for the wild type, pI = 5.0; for the K234E mutant, pI = 4.8. The decrease of 0.2 in the pI is expected, based on the change from the positively charged ammonium group of Lys-234 to the negatively charged carboxylate of Glu-234. Therefore, effects on catalysis do not stem from gross changes in the protein's conformation.

The binding affinity of the enzyme for substrate can be determined from the ES dissociation constant, K_s. Values of K_s were estimated from the observed kinetics on the following basis. For the acyl-enzyme kinetic scheme, $K_m = k_3 K_s/(k_2 + k_3)$, where k_2 and k_3 are the rate constants for acylation and deacylation, respectively. Thus, if acylation is rate limiting, $K_m = K_s$, whereas if deacylation is rate limiting $K_m < K_s$. In order to determine which step was rate limiting, we used subzero temperature trapping of the acyl enzyme using FAP and ^{14}C-labeled benzylpenicillin as substrates. Since the covalent intermediate could be trapped, the results indicate rate-limiting deacylation.

The major effects of the mutation on catalysis were (1) a pH-dependent increase in K_m, ranging from a negligible effect at low pH to two orders of magnitude at high pH (Figure 3, Table 1), (2) a decrease in k_{cat} on the order of two orders of magnitude (Figure 4, Table 1) as well as a (3) substantial

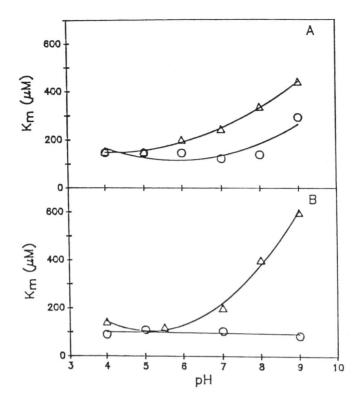

FIGURE 3. The effect of pH on K_m for the β-lactamase-catalyzed hydrolysis of phenoxymethylpenicillin. Data for the wild-type enzyme are shown by ○, those for K234E by △. The ionic strength was 0.15 M; temperature was 30°C. (Reprinted with permission from Ellerby, L. M., Escobar, W. A., Fink, A. L., Mitchinson, C., and Wells, J. A., *Biochemistry*, 29, 5797, 1990. Copyright 1990, American Chemical Society.)

shift in pK_1 for k_{cat} (the shape of the pH-rate profiles are very sensitive to ionic strength; at 0.5 M, the mutant profile is much more similar to that of the wild type), (4) an increase in the acidic pK of the bell-shaped curve for k_{cat}/K_m (e.g., for phenoxymethylpenicillin, pK_1 increased from 5.0 to 5.7) (Figure 5) and no change in the alkaline pK, reflecting effects on the pKs of the essential catalytic groups in the free enzyme (Figure 5), and (5) small differences in the substrate specificity profile (see Table 1).

If we assume that the major role of Lys-234 is on substrate binding, then we would expect the effects of the K234E mutation to be most clearly manifest in the K_s and K_m terms, i.e., ground state binding. What is observed is not only such an effect, but also a substantial decrease in k_{cat}, reflecting a decrease in transition state binding. As a consequence, the catalytic efficiency of the mutant is decreased three to four orders of magnitude at some pHs. As might be anticipated, the conversion of a positively charged side chain into a negatively

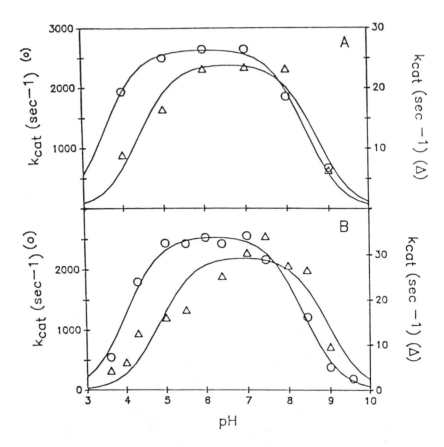

FIGURE 4.   The effect of pH on k_{cat} for the β-lactamase-catalyzed hydrolysis of phenoxymethylpenicillin. The wild-type data are shown by ○, those for K234E by △. The ionic strength
was 0.15 *M*; temperature was 30°C.

TABLE 1
Substrate Specificity Profiles of Wild-Type β-Lactamase from *B. licheniformis* and the K234E Mutant

	Wild-type enzyme			Mutant		
Substrate	k_{cat}	K_m	k_{cat}/K_m	k_{cat}	K_m	k_{cat}/K_m
Benzylpenicillin	2650	123	2.15×10^7	49.6	8100	6.1×10^3
Phenoxymethylpenicillin	2552	103	2.48×10^7	34.3	1800	1.9×10^4
Ampicillin	2241	211	1.06×10^7	21.0	13700	1.5×10^3
Nitrocefin	1088	41.2	2.64×10^7	7.9	1220	6.5×10^3

Note: Units were s^{-1} for k_{cat}, μM for K_m, and $M\ s^{-1}$ for k_{cat}/K_m. Conditions were pH 7.0,
0.15 *M* KCl, 30°C.

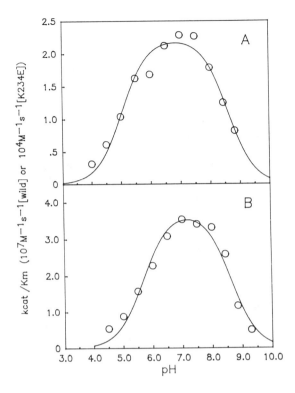

FIGURE 5. The effect of pH on k_{cat}/K_m for phenoxymethylpenicillin at ionic strength 0.5 M. (A) Data for K234E; (B) data
for wild-type β-lactamase. (Reprinted with permission from
Fink, A. L., Ellerby, L. M., and Bassett, P. M., *Biochemistry*,
111, 6871, 1989. Copyright 1989, American Chemical Society.)

charged one results in large changes in the pKs of the active-site groups.
Interestingly, there are much more significant perturbations of the pKs of the
ES complex (as seen in the large increase in pK_1 and pK_2 for k_{cat}) than in the
pKs of the essential catalytic groups in the free enzyme (as reflected by the
pKs for k_{cat}/K_m).

How can our observations be accounted for, based on the known structure
and properties of β-lactamase and its potential mechanism of action? Possi-
bilities include the following. (1) In the K234E mutant, a new interaction is
formed between the carboxylate of Glu-234 and some nearby cationic residue,
which is, of course, absent in the wild-type enzyme, and which leads to both
a change in local structure in the vicinity of residue 234 and to a decrease in
the effective charge on the side chain of Glu-234. This possibility can be
eliminated on the basis of the lack of observed structural perturbation in the
mutant and examination of the structure of the enzyme, which shows no suitable residues to form such a salt bridge in the neighborhood of residue 234.

(2) The carboxylate of Glu-234 may be sufficiently buried in the relatively hydrophobic floor of the active site, and presumably hydrogen bonded to water molecules, that the unfavorable interaction between the C3 carboxylate of the penicillin substrate and the Glu-234 side chain is significantly decreased. Water has a high dielectric constant, and the difference in length between lysine and glutamate side chains make this a reasonable assumption. In addition, Hwang and Warshel[28] have shown, using free-energy perturbation calculations for the aspartate aminotransferase system,[15] that the local environment in the enzyme is likely to result in substantially different free energies of interaction between a pair of charged groups, depending on the relative positions. However, this explanation should still result in substantially decreased binding, due to the loss of the favorable electrostatic interaction with Lys-234. The observed pH dependence of K_m is consistent with such an interpretation. (3) The presence of the negatively charged Glu-234 leads to repulsion of the substrate carboxylate, leading to an alternative orientation of the substrate in the active site. The affinity for the ground state of the substrate is not drastically decreased compared to that with the wild-type enzyme, but the different orientation means that the relative spatial location of the catalytic groups is less optimal and hence catalysis is decreased. Thus, the ground state binding affinity is decreased somewhat in the mutant due to alternative compensating binding interactions between substrate and enzyme, and the altered spatial orientation of the substrate relative to the catalytic groups leads to weaker binding of the transition state, and hence lower k_{cat}. We believe this last explanation to be the best at the present time.

The results with the K234E mutant reveal that β-lactamase is better able to compensate for such a mutation than other enzymes in related experiments in which the charge of an enzyme group involved in an interaction with substrate was reversed. For example, with trypsin, Graf et al.[25] replaced Asp-189, which determines the substrate specificity, with a Lys residue. It had been assumed that an electrostatic interaction with the side chain of Lys or Arg in the substrate resulted in proper orientation of the substrate with respect to the catalytic residues. In contrast to the case with β-lactamase, the resulting mutant trypsin was reported to be totally inactive toward Lys and Arg substrates. In addition, the trypsin mutant did not exhibit compensatory activity toward corresponding Asp or Glu substrates. Similarly, Cronin and Kirsch[15] converted the specificity-determining Arg-292 in aspartate aminotransferase into Asp. Again, in contrast to the case of β-lactamase, the activity of the aspartate aminotransferase mutant toward the natural substrates was depressed by five orders of magnitude, and the K_m values of the mutant were much greater than those of the wild type and too large to allow determination. However, the results obtained with β-lactamase are more similar to those observed with subtilisin, in which mutations involving a change from Gln-156/Lys-166 to Gln-156/Asp-166 in the specificity pocket of subtilisin resulted in a three orders of magnitude decrease in k_{cat}/K_m and a corresponding increase in K_m.[43]

VI. CATALYTIC MECHANISM FOR β-LACTAMASE

Given the strongly conserved amino acid sequence homology between β-lactamase from *S. aureus* and *B. licheniformis* and the similarity in three-dimensional structure,[30] it is apparent that the only charged groups likely to be directly involved in catalysis are Lys-73 and -234, and Glu-166. Since the pH dependence of k_{cat}/K_m of both wild-type and K234E mutant are essentially the same, we conclude that the groups responsible are Glu-166 for pK_1 and Lys-73 for pK_2. The present results indicate that, although conserved, Lys-234 is not an essential group for binding or catalysis. Although k_{cat} is depressed in the mutant, the mutant is still a very good catalyst (see Table 1). The effects of the K234E mutation on the ionizing groups responsible for the pH dependence of k_{cat}, K_m, and k_{cat}/K_m are readily explained as being due to the expected electrostatic effects of the replacement of an ammonium group by a carboxylate in the general vicinity of the active site.

We propose that catalysis by β-lactamase involves initial substrate binding in which positively charged Lys-234 facilitates orientation of the substrate by electrostatic interactions with the substrate carboxylate followed by nucleophilic attack by Ser-70, aided by general base catalysis by Glu-166 (Scheme 1). The resulting tetrahedral adduct is stabilized by hydrogen bonds between the oxyanion and main chain amides.[26] Collapse of the tetrahedral adduct involves Lys-73 acting as a general acid catalyst to donate a proton to the lactam N, leading to formation of the acyl-enzyme intermediate. Deacylation involves the free-base form of Lys-73 acting as a general base to activate water to attack the acyl carbonyl. Breakdown of the resulting tetrahedral adduct involves Glu-166 acting as a general acid and restores the enzyme to its original state ($Glu-COO^-$ and $Lys-NH_3^+$). The essentiality of Glu-166[34] and Lys-73 have been previously reported, although if the latter is replaced by Arg, significant activity remains.[42]

One can speculate that the normal role of Lys-234 is electrostatic facilitation of substrate binding, but in the K234E mutant this role can be successfully taken over by Lys-73 and H-bonding by one or more water molecule(s) in the space left by the replacement of Lys by Glu. Helix dipole moments may also contribute in this regard.[26] As noted, we assume that in the mutant the mode of substrate binding is perturbed from that of the wild type. It is possible that the poorer binding of the transition state in the mutant results from the absence of the electrostatic interaction to the ammonium group of Lys-234.

VII. MECHANISM OF INHIBITION BY PENICILLIN SULFONES

Although few classes of inhibitors for the β-lactamases are known, there is considerable interest in such compounds because of their potential clinical importance. Penicillanic acid sulfone was made semisynthetically by English et al.[18]

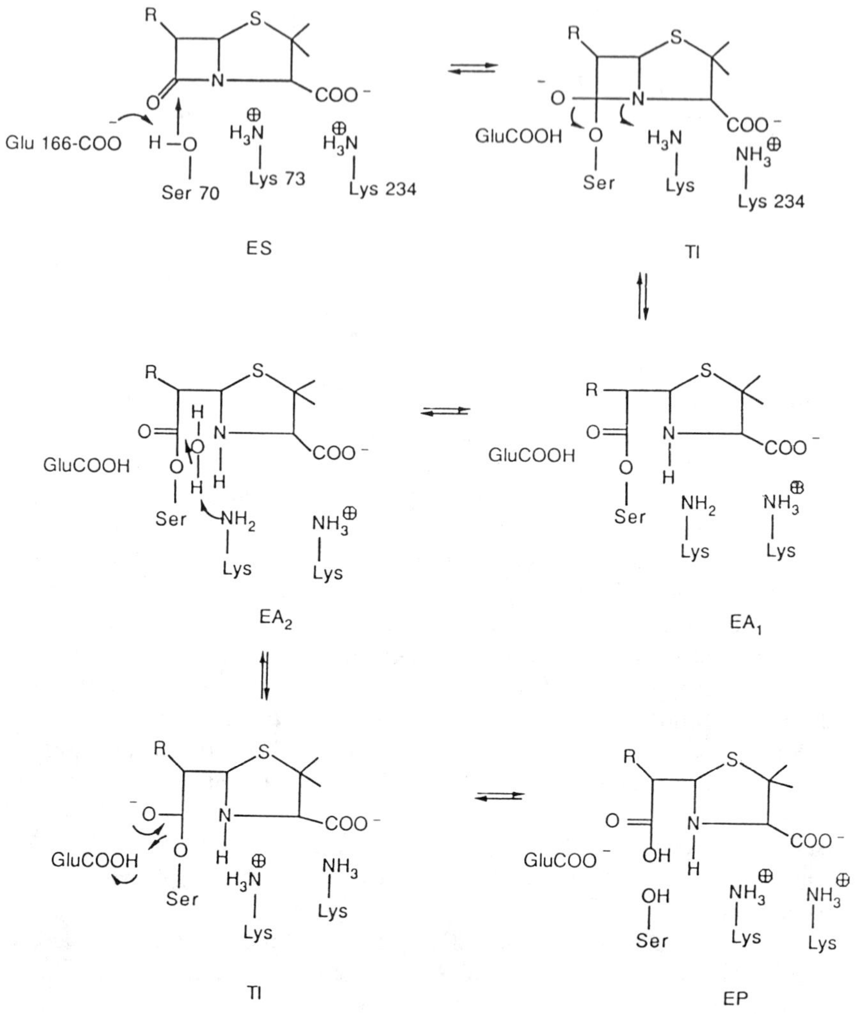

SCHEME 1.

and shown to be a good inhibitor of β-lactamases. Subsequently, the sulfones of several other penicillins have been made and shown to be inhibitors.[6,13,17,23]

The inhibition of β-lactamases by penam sulfones appears to be a form of suicide inactivation in which the lability of the C-S bond results in the formation of a transient imine acyl-enzyme intermediate which tautomerizes to the more stable enamine form. This tautomerization is driven by conjugation with the ester carbonyl, leading to a vinylogous ester of reduced hydrolytic sensitivity. The enamine, which is also a β-amino acrylate, is characterized by a strongly absorbing chromophore in the 280 to 320 nm region. Previous studies indicate that inactivation of β-lactamase by quinacillin sulfone and by 6-β-(trifluoromethanesulfonyl)amidopenicillanic acid sulfone resulted in covalent modification of the active-site serine.[22,27]

The inhibitory reaction has been studied in the most detail with penicillanic acid sulfone, especially by Knowles and co-workers.[3,4,31] They conclude that the acyl enzyme can undergo three fates: (1) hydrolysis to give turnover, (2) conversion to the enamine form of the acyl enzyme, leading to transient inhibition, and (3) transimination by a suitably positioned lysine residue in the active site, leading to irreversible inactivation.[3] The low pK of the sulfinate leaving group, the acidity of the C6 proton (removed in the formation of the enamine), the relative stability toward hydrolysis of vinylogous esters, and the potential for transimination all provide a good *chemical* rationale for the observed inhibitory effects of penam sulfones on β-lactamase.

In an examination of sulfone-inactivated β-lactamase, we noted a number of properties which suggested that inactivation resulted in the exposure of substantial hydrophobic surface area, which indicated that the conformation of the inactivated enzyme was different from that of the native state. Dmitrienko et al.[17] have reported that inactivation of β-lactamase by 6-β-(trifluoromethanesulfonyl)amidopenicillanic acid sulfone leads to a species which differs from the native enzyme in terms of its thermostability and optical rotary dispersion spectrum. We have undertaken a more detailed investigation of the role of conformational processes in the reversible and irreversible inactivation of β-lactamase by penam sulfones in order to ascertain the contributions of chemical and physical processes to the inhibition. In the experiments discussed here, we used circular dichroism to examine the role of secondary structure changes in the inactivation of the enzyme.

We used β-lactamase I from *B. cereus* and the sulfones of nafcillin, cloxacillin, and penicillanic acid. The former sulfones are derived from Type A substrates, which bring about substrate-induced deactivation.[12,35] Nafcillin sulfone is a very efficient inactivator of β-lactamase; the ratio of turnover to inactivation is less than 5:1 at pH 9 and temperatures of 37°C or above. This is the smallest such ratio reported for a penam sulfone.

β-Lactamase (5 μM) was inactivated at pH 9.0, at either 30 or 37°C with the Type A sulfones with [S]/[E] = 20:1 or 100:1. Inactivation was monitored by removing aliquots and diluting them into a benzylpenicillin assay. Essentially complete inactivation occurs in 2 h or less under these conditions. Under appropriate conditions, both reversible and irreversible inhibition were noted during the early stages of inactivation. The inactivated enzyme was stable for long periods (days at 4°C) at alkaline pH, but can be completely reactivated on acidifying the solution to pH < 7, due to an acid-catalyzed process (probably imine hydrolysis). The far-UV circular dichroism spectra of β-lactamase inactivated by the sulfones of nafcillin and cloxacillin were essentially identical, and are shown in Figure 6A along with that of the native enzyme (pH 9.0). Reactivation of the sulfone-inactivated enzyme by dropping the pH to 6.0 for 2 h leads to a return of the native spectrum. Inactivation leads to

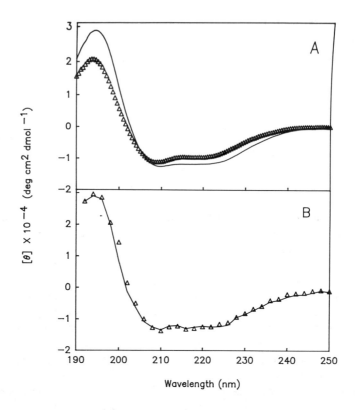

FIGURE 6.   The Far-UV circular dichroism spectra of native and sulfone-inactivated β-lactamase. (A) Native (−) and nafcillin sulfone-inactivated enzyme (△). Conditions were pH 9.0, 4°C, [E] = 5.7 μ*M*, [S] = 120 μ*M*. (B) Native (−) and penicillanic acid sulfone-inactivated enzyme (△). Conditions were pH 8.9, 4°C, [E] = 5.7 μ*M*. The inactivation was carried out at 30°C with [penicillanic acid sulfone] = 8.6 m*M*.

spectral changes consistent with the loss of α-helix (8% by deconvolution). We therefore conclude that inactivation by the sulfones of Type A substrates leads to a change in the secondary structure of the enzyme. The inactivated enzyme is in the form of an enamine, based on the development of a chromophore at 295 nm during inactivation. Under the above experimental conditions, the contribution of nafcillin sulfone or cloxacillin sulfone or their products to the far-UV circular dichroism spectrum is negligible.

The kinetics of the inactivation reaction were compared with the change in ellipticity at 196 nm in an experiment with nafcillin sulfone using 5.7 μ*M* enzyme (the rates are proportional to the enzyme concentration), [S]/[E] = 20:1, 36.8°C. The observed first-order rate constants were identical within experimental error, namely, 6.9×10^{-4} and 7.3×10^{-4} s^{-1}, respectively. Thus, under these experimental conditions, the conformational change and the inactivation process occur simultaneously.

The inactivation was also monitored by tryptophan fluorescence. The fluorescence emission of the enzyme was quenched in a time-dependent manner, with a first-order rate constant identical to that observed for inactivation $(7.3 \times 10^{-4} \text{ s}^{-1})$. Since the three tryptophan residues are located far from the active site, we conclude that the quenching is due to a conformational change in the protein in the vicinity of the Trp residues.

Interestingly, β-lactamase irreversibly inactivated by the sulfones of the Type A substrates shows many conformational properties similar to those of the noncatalytically active A or B states.[24] The latter are found at acidic or alkaline pH and high ionic strength, and have properties consistent with a molten globule state (i.e., compact, substantial secondary structure, but disordered tertiary structure). Properties in common include similar far-UV circular dichroism spectra, 1-anilino-naphthalene-8-sulfonic acid (ANS) binding, and similar quenched tryptophan fluorescence; however, the sulfone-inactivated state appears to be similar in compactness to the native state, in contrast to the A and B states, which are more expanded.[24]

In contrast to the case of penam sulfones derived from Type A substrates which give *irreversibly* inactivated enzyme at high pH, the penicillanic acid sulfone-inactivated enzyme was only *reversibly* inactivated at high pH (even at [S]/[E] of several thousand). β-lactamase (5.6 μ*M*) was inactivated with penicillanic acid sulfone at pH 8.9 with [S]/[E] = 3,000:1.[22] The large concentration of sulfone necessitated its removal prior to collecting the circular dichroism data. This was accomplished by using two passages through a centrifugal mini ion-exchange column. The separation was done at 1°C. Due to the significant rate of reactivation under these conditions, some 20 to 25% of the inactivated enzyme had reactivated by the time the first spectrum was collected. Spectra were collected at pH 8.9 (at 4°C) at frequent time periods during the reactivation process, which was monitored simultaneously. As shown in Figure 6B, the spectrum was identical to that of the native enzyme at all times. We therefore conclude that for penicillanic acid sulfone-inactivated enzyme, there was no significant change in the secondary structure due to inactivation.

When the Trp fluorescence was followed during inactivation with penicillanic acid sulfone, quenching was observed, the rate being the same as that of inactivation. Thus, it is likely that a change in the tertiary structure occurred with the same rate as inactivation. Reactivation was accompanied by an increase in Trp fluorescence emission, with the same rate as reactivation $(3.2 \times 10^{-3} \text{ s}^{-1}, 4°C, \text{ pH } 9.0)$.

These observations illustrate that there are differences in the mechanism of inactivation of penam sulfones due to the nature of the side chain on C6, especially in the case of those in which the parent compounds are Type A substrates. In addition, it appears that the combination of the flexibility introduced into the inhibitor by conversion from a fused bicyclic ring system to an acyclic one in the enamine form of the acyl enzyme, as well as the inherent mobility of the enzyme, leads to a significant conformational change on formation of the enamine acyl enzyme. Based on the behavior of enzyme inactivated by penicillanic acid sulfone, which lacks a side chain on C6 and shows neither the change in secondary structure nor stability at alkaline pH, we believe it is the conformational change, rather than the slow hydrolysis of the enamine, which leads to the irreversible inactivation at alkaline pH by the sulfones of Type A substrates. This conformational change results in displacement of the catalytic groups with respect to the acyl-enzyme bond such that catalysis of deacylation is prevented.

Under our experimental conditions, penicillanic acid sulfone is not an irreversible inactivator, and the observed reversible inhibition is presumably due to a combination of the lower reactivity of the enamine form of the acyl enzyme to hydrolysis (i.e., enzyme-catalyzed deacylation) and possibly the effects of a change in the tertiary structure.

ACKNOWLEDGMENTS

The influence of Myron Bender's approach to studying enzyme mechanisms should be very apparent in this work. It is a great pleasure to acknowledge his contributions to my own development as a scientist and to the field of enzyme chemistry. This research was supported by grants from The National Institutes of Health and The National Science Foundation. I am very grateful to the assistance of my co-workers in this research — Pat Bassett, Ed Chichester, Lisa Ellerby, Walter Escobar, Yuji Goto, Colin Mitchinson, Sally Swedberg, Tony Tan, Richard Virden, and Jim Wells — without whose contributions this work would not have been possible.

REFERENCES

1. **Anderson, E. G. and Pratt, R. F.,** *J. Biol. Chem.,* 256, 11401, 1981.
2. **Anderson, E. G. and Pratt, R. F.,** *J. Biol. Chem.,* 258, 13120, 1983.
3. **Brenner, D. G. and Knowles, J. R.,** *Biochemistry,* 20, 3680, 1981.
4. **Brenner, D. G. and Knowles, J. R.,** *Biochemistry,* 23, 5833, 1984.
5. **Carrey, E. A. and Pain, R. H.,** *Biochim. Biophys. Acta,* 533, 12, 1978.
6. **Cartwright, S. J. and Coulson, A. F. W.,** *Nature,* 278, 360, 1979.
7. **Cartwright, S. J. and Coulson, A. F. W.,** *Philos. Trans. R. Soc. London Ser. B,* 289, 370, 1980.
8. **Cartwright, S. J. and Fink, A. L.,** *CRC Crit. Rev. Biochem.,* 11, 145, 1981.
9. **Cartwright, S. J. and Fink, A. L.,** *FEBS Lett.,* 137, 186, 1982.
10. **Cartwright, S. J., Tan, A. K., and Fink, A. L.,** *Biochem. J.,* 263, 905, 1989.
11. **Christenson, J. G., Pruess, D. L., Talbot, M. K., and Keith, D. D.,** *Antimicrob. Agents Chemother.,* 32, 1005, 1988.
12. **Citri, N., Samuni, A., and Zyk, N.,** *Proc. Natl. Acad. Sci. U.S.A.,* 73, 1048, 1976.
13. **Clarke, A. J., Mezes, P. S. F., Vice, S. F., Dmitrienko, G. I., and Viswanatha, T.,** *Biochim. Biophys. Acta,* 748, 389, 1983.

14. **Cohen, S. A. and Pratt, R. F.,** *Biochemistry,* 19, 3996, 1980.
15. **Cronin, C. N. and Kirsch, J. F.,** *Biochemistry,* 27, 4572, 1988.
16. **Dalbadie-McFarland, G., Cohen, L. W., Riggs, A. D., Itakura, K., and Richards, J. H.,** *Proc. Natl. Acad. Sci. U.S.A.,* 79, 6409, 1982.
17. **Dmitrienko, G. I., Copeland, C. R., Arnold, L., Savard, M. E., Clarke, A. J., and Viswanatha, T.,** *Bioorg. Chem.,* 13, 34, 1985.
18. **English, A. R., Retsema, J. A., Girard, A. E., Lynch, J. E., and Barth, W. E.,** *Antimicrob. Agents Chemother.,* 14, 414, 1978.
19. **Fink, A. L. and Geeves, M. A.,** *Methods Enzymol.,* 63A, 336, 1979.
20. **Fink, A. L. and Petsko, G. A.,** *Adv. Enzymol.,* 52, 177, 1981.
21. **Fink, A. L., Behner, K. M., and Tan, A. K.,** *Biochemistry,* 26, 4248, 1987.
22. **Fisher, J., Charnas, R. L., and Knowles, J.,** *Biochemistry,* 17, 2180, 1978.
23. **Fisher, J., Charnas, R. L., Bradley, S. M., and Knowles, J. R.,** *Biochemistry,* 20, 2726, 1981.
24. **Goto, Y. and Fink, A. L.,** *Biochemistry,* 28, 945, 1989.
25. **Graf, L., Craik, C. S., Patthy, A., Roczniak, S., Fletterick, R. J., and Rutter, W. J.,** *Biochemistry,* 26, 2616, 1987.
26. **Herzberg, O. and Moult, J.,** *Science,* 236, 694, 1987.
27. **Hilhorst, I. M., Dmitrienko, G. I., Viswanatha, T., and Lampen, J. O.,** *J. Prot. Chem.,* 3, 275, 1984.
28. **Hwang, J.-K. and Warshel, A.,** *Nature,* 334, 270, 1988.
29. **Joris, B., Ghuysen, J.-M., Dive, G., Renard, A., Dideberg, O., Charlier, P., Frere, J.-M., Kelly, J. A., and Boyington, J. C.,** *Biochem. J.,* 250, 313, 1988.
30. **Kelly, J. A., Dideberg, O., Charlier, P., Wery, J. P., Libert, M., Moews, P. C., Knox, J. R., Duez, C., Fraipont, C., Joris, B., Dusart, J., Frere, J.-M., and Ghuysen, J.-M.,** *Science,* 231, 1429, 1986.
31. **Kemal, C. and Knowles, J. R.,** *Biochemistry,* 20, 3688, 1981.
32. **Knott-Hunziker, V., Waley, S. G., Orlek, B. S., and Sammes, P. G.,** *FEBS Lett.,* 99, 56, 1979.
33. **Kuwabara, S. and Abraham, E. P.,** *Biochem. J.,* 103, 27c, 1967.
34. **Madgwick, P. J. and Waley, S. G.,** *Biochem. J.,* 248, 657, 1987.
35. **Pain, R. H. and Virden, R.,** in *Beta-Lactamases,* Hamilton-Miller, J. M. T. and Smith, J. T., Eds., Academic Press, New York, 1979, 141.
36. **Sigal, I. S., Harwood, B. G., and Arentzen, R.,** *Proc. Natl. Acad. Sci. U.S.A.,* 79, 7157, 1982.
37. **Sigal, I. S., Degrado, W. F., Thomas, B. J., and Petteway, S. R.,** *J. Biol. Chem.,* 259, 5327, 1984.
38. **Van Iersel, J., Jzn, J. F., and Duine, J. A.,** *Anal. Biochem.,* 151, 196, 1985.
39. **Varetto, L., Frere, J.-M., and Ghuysen, J.-M.,** *FEBS Lett.,* 225, 218, 1987.
40. **Virden, R., Tan, A. K., and Fink, A. L.,** *Biochemistry,* 29, 145, 1989.
41. **Joy, D. and Fink, A. L.,** unpublished results.
42. **Richards, J. H.,** personal communication.
43. **Wells, J. A., Powers, D. B., Bott, R. R., Graycar, T. P., and Estell, D. A.,** *Proc. Natl. Acad. Sci. U.S.A.,* 84, 1219, 1987.

Chapter 5

EFFECT OF ASCORBIC ACID AND COPPER ON PROTEINS

**Avi Golan-Goldhirsh, David T. Osuga, Andi O. Chen, and
John R. Whitaker**

TABLE OF CONTENTS

I. INTRODUCTION

A. RELATIONSHIP BETWEEN ASCORBIC ACID AND COPPER

Despite extensive studies on the role of ascorbic acid in preventing scurvy and its possible function with 11 monooxygenase and dioxygenase systems,[1] the exact role of ascorbic acid in human metabolism in relation to other reductants cannot be stated with certainty. In all 11 of the enzyme systems mentioned, other reductants will replace ascorbic acid in *in vitro* systems, but not as efficiently. It is of interest that ascorbic acid is associated with 11 enzyme systems where the oxidation state of copper shuttles between the Cu^+ and Cu^{2+} states and iron between the Fe^{2+} and Fe^{3+} states. Does ascorbic acid perform only the role of reducing these metal ions at the end of a catalytic cycle?

The major consequences of ascorbic acid and copper (or iron), alone or combined, in an *in vitro* system are just now becoming well known. This paper will be devoted to this topic.

B. COPPER-CATALYZED REACTIONS IN *IN VITRO* SYSTEMS

Several recent publications have described the effect of added copper in enhancing the rates of several biological reactions. Samuni et al.[2] reported on the unusual copper-induced sensitization of biological damage due to superoxide radicals. The model used was the response of penicillinase to radiation.

Yamauchi and Seki[3] found that Cu^{2+} catalyzed the disulfide bond cleavage of *bis*{2-[(2-pyridylmethyl)amino]ethyl}disulfide, forming 2-[(2-pyridylmethyl)amino]ethylsulfinate. The reaction only occurred in the presence of O_2. Florence et al.[4] reported that the toxic effect of 2,9-dimethyl-1,10-phenanthroline, accumulated by bacteria, fungi, mycoplasmas, tumor cells, algae, and amphipods, is a result of its complexation with Cu^+, formed by the reduction of Cu^{2+} by H_2O_2. However, higher concentrations of H_2O_2 that accumulated in the organisms resulted in oxidation of the phenanthroline ring and loss of toxicity. Cu^{2+} in the presence of imidazole and O_2 efficiently oxidized (86% yield) anthracenols and anthrone to anthraquinones.[5]

Jones et al.[6] reported that Cu^{2+}, in the presence of −SH groups of cell membranes and energy from light of >425 nm, forms Cu^+. Cu^+ in turn reacts with O_2 to form a superoxide radical (O_2^-), which can cause peroxidation of membrane lipids. Lipsky and Ziff[7,8] found that copper increases the effect of D-penicillamine on nitrogen-induced human lymphocyte proliferation, probably by induction of H_2O_2 formation; this postulation has been supported by more recent research.[9] The enhancement was eliminated by addition of catalase to the reaction.

Sporidesmin, the mycotoxin responsible for "facial eczema" in ruminants, is known to form superoxide radicals as a result of the autoxidation of the reduced form of the mycotoxin, a dithiol. Addition of Cu^{2+} increased the rate of autoxidation several hundred-fold.[10] It was suggested that superoxide radicals were produced.

Treatment of bovine mitochondrial H^+-ATPase with copper-o-phenanthroline, known to be an SH-oxidizing cross-linking reagent, resulted in cross linking of the enzyme via a disulfide bond, with loss of activity.[11] The reactions were performed in the air at 23°C for 2 min. Cooper et al.[12] showed that added Cu^{2+} greatly enhanced the limited degradation of mucus glycoproteins by H_2O_2. They suggested that specificity of the attack at the histidine residues was a result of the binding of Cu^{2+} or Fe^{2+} at these residues only.

C. ASCORBIC ACID-COPPER-CATALYZED REACTIONS

Ascorbic acid is added to foods in substantial quantities to prevent the formation of N-nitroso compounds in nitrite-preserved meats[13] and to prevent enzymatic and nonenzymatic browning reactions. In preventing N-nitroso compound formation, the reaction of ascorbic acid at pH 3 to 4 is thought to involve the reduction of NO_2 to NO, as shown in Equation 1.[13]

$$\text{ascorbic acid} + 2HNO_2 \rightarrow \text{dehydroascorbic acid} + 2NO + H_2O \quad (1)$$

Carbohydrates — Ascorbic acid-copper ion has been reported to cause the oxidative degradation of β-cyclodextrin.[14] The authors suggested that the degradation resulted from the formation of ·OH radicals generated by the autoxidation of ascorbic acid in the presence of Cu^{2+}. Ascorbic acid alone, or in the presence of copper or iron ions, depolymerized hyaluronic acid.[15] The rate of the ascorbic acid-catalyzed reaction (0.33 mM ascorbic acid) was increased 1.7 and 3.3 times in the presence of $FeSO_4$ and $CuSO_4$, respectively.

Nucleic Acids — Murata et al.[16] reported that bacteriophage J1 was inactivated by ascorbic acid. It was shown that bubbling air and the addition of oxidizing agents and Cu^{2+} enhanced the rate, while bubbling N_2 or the addition of reducing agents or radical scavengers prevented the inactivation.[17] In 1976, Stich et al.[18] reported that "oxidized" ascorbic acid and ascorbic acid-Cu^{2+} caused DNA fragmentation. $CuSO_4$ solutions alone, ranging from 10^{-6} to 10^{-4} M, had no detectable effect on DNA and chromosomes. There was no effect when the reaction was performed in N_2. Ascorbic acid was shown to inhibit the growth of mouse neuroblastoma and human endometrial carcinoma cells at concentrations greater than 100 μM.[19] This antitumor activity of ascorbic acid was greatly enhanced by adding copper ions or copper chelates. The effect was shown to be due to specific cleavage of DNA.[20]

Proteins — Prevention of enzymatic browning by ascorbate is generally thought to involve reduction of the product, a benzoquinone, to the *o*-dihydroxyphenol, the substrate of the enzyme reaction, with the oxidation of ascorbate to dehydroascorbate[21] (Equation 2).

However, incubation of polyphenol oxidase alone with ascorbate led to loss of enzymatic activity.[21] The shape of the product vs. time curve indicated that an intermediate of ascorbate oxidation might be involved in the inactivation of polyphenol oxidase.[22,23] Dehydroascorbate was more effective than ascorbate in inactivating polyphenol oxidase. Incubation of ascorbate in air prior to adding it to polyphenol oxidase led to an even more rapid loss of enzyme activity. Ascorbate in the presence of added Cu^{2+} was much more effective than either ascorbate or dehydroascorbate, and the source of the Cu^{2+} made no difference. At a fixed ascorbate concentration of 5 mM, the rate of loss of polyphenol oxidase activity was dependent on the Cu^{2+} concentration, up to 20 μM. There was no further effect of Cu^{2+} concentration on the rate of activity loss between 20 and 790 μM Cu^{2+}, suggesting a saturation effect. The rate of loss of enzyme activity was also dependent on the ascorbate concentration, but did not show a saturation effect with respect to ascorbate. O_2 was required for the reaction.

II. EXPERIMENTAL

A. MATERIALS
Mushroom polyphenol oxidase (PPO; grade III, lot 91F-9650) and bovine serum albumin (BSA) were from Sigma Chemical Co. Soybean trypsin inhibitor was from Nutritional Biochemical Corp. Ovalbumin was a gift (prepared by Dave Osuga of R. E. Feeney's laboratory). L-Ascorbic acid was from Aldrich Chemical Co. and pyrocatechol was from Eastman Organic Chemicals. Other chemicals used were of analytical grade. Deionized distilled water was used.

B. METHODS
Enzyme Assays — The spectrophotometric assay for polyphenol oxidase activity was carried out on a Varian 635 or Beckman 35 spectrophotometer equipped with a thermostated cuvette holder and a recorder. The reaction, final volume of 3 ml, contained air-saturated 0.1 M sodium phosphate buffer,

pH 6.5, 10 m*M* pyrocatechol, and 7 μg/ml mushroom PPO at 30°C. The reaction was started by the addition of enzyme. The polarographic assay was carried out on a YSI Model 53 biological oxygen monitor. Change in oxygen concentration was recorded continuously on a Honeywell recorder attached to the oxygraph. The percent remaining activity was calculated as the activity of a treated sample compared to that of a reference sample treated identically, but not including ascorbate, Cu^{2+}, or ascorbate-Cu^{2+}.

Polyacrylamide Gel Electrophoresis — Nondenaturing polyacrylamide gel electrophoresis (PAGE) and sodium dodecyl sulfate (SDS)-polyacrylamide gel electrophoresis (SDS-PAGE) were performed according to the *Hoefer Scientific Instrument Manual*,[24] with some modifications.

Size-Exclusion Chromatography — Polyphenol oxidase, bovine serum albumin, soybean trypsin inhibitor, and ovalbumin were chromatographed on Sephadex G-50 (1.8 × 39.5 cm) at pH 6.5 and 4°C in 0.05 *M* phosphate buffer-0.1 *M* NaCl containing 0.02% NaN_3. The elution patterns were monitored at 280 nm.

Protein Treatment with Ascorbate, Cu^{2+}, or Ascorbate-Cu^{2+} — Protein (19 μ*M* usually) was incubated with 5 m*M* ascorbic acid, 0.8 m*M* cupric sulfate, or 5 m*M* ascorbic acid-0.8 m*M* cupric sulfate in 0.1 *M* sodium phosphate buffer, pH 6.5, at 25°C for 22 h in a gyrotory water bath shaker.

Protein Hydrolysis and Amino Acid Analysis — The proteins were hydrolyzed for 24 h in 6 *M* constant boiling HCl in sealed Pyrex tubes at 110°C. The HCl was removed by repeated vacuum distillation and the hydrolyzate dissolved in the appropriate buffer. Amino acid analyses were performed on an automated Durrum D-500 instrument.

III. RESULTS

A. LOSS OF POLYPHENOL OXIDASE ACTIVITY

As shown in Table 1, incubation of PPO under aerobic conditions at pH 6.5 and 25°C for 60 min indicated it was completely stable. When PPO was incubated under the same conditions, except that 5 m*M* ascorbic acid and 0.80 m*M* $CuSO_4$ were added, 72% of the activity was lost when incubation was under aerobic conditions, but 0% activity was lost under anaerobic conditions (N_2) after 60 min. On restoration of aerobic conditions, 70% of the activity was lost in 54 min.

B. AMINO ACID COMPOSITION OF PROTEINS BEFORE AND AFTER TREATMENT WITH ASCORBATE, Cu^{2+}, OR ASCORBATE-Cu^{2+}

The amino acid compositions of the proteins before and after treatment with ascorbate, Cu^{2+}, or ascorbate-Cu^{2+} were determined. There were changes in the amino acid composition of polyphenol oxidase following treatment with ascorbate-Cu^{2+} (Table 2). In particular, on a subunit basis, there was a decrease of histidine (from 4.2 to 1.1) and methionine (from 1.5 to 0.9) residues, and increases in aspartic acid/asparagine (from 27.4 to 33.9),

TABLE 1
Requirement for Oxygen in the Inactivation of Polyphenol Oxidase (PPO) by Ascorbate-Cu^{2+}

Condition[a]	Incubation time (min)	Activity remaining (%)
Aerobic		
Buffer + PPO	1	100
	60	100
Buffer + PPO + ascorbic acid-Cu^{2+} [b]	1	100
	60	28
Anaerobic[c]		
Buffer + PPO + ascorbic acid-Cu^{2+}	1	100
	60	100
	114[d]	30

[a]   7.1 μM polyphenol oxidase in 0.11 M sodium phosphate buffer, pH 6.5, and 25°C.
[b]   5 mM ascorbic and 0.80 mM CuSO$_4$.
[c]   All solutions were deaerated thoroughly by vacuum before mixing and the system was continuously flushed with N$_2$.
[d]   After 60 min under N$_2$, aerobic conditions restored.

TABLE 2
Amino Acid Composition[a] of Polyphenol Oxidase Incubated with Ascorbate, Cupric Sulfate, or Ascorbate-Cupric Sulfate

Amino acid	Untreated	Ascorbate treated[b]	Cupric sulfate treated[c]	Ascorbate-Cu^{2+} treated[d]
Asx	27.4	27.8	26.9	33.9
Thr	15.9	15.9	16.2	18.2
Ser	14.7	15.0	14.1	16.4
Glx	26.9	27.0	26.4	31.2
Pro	11.8	11.8	12.0	11.7
Gly	19.9	20.9	20.5	23.3
Ala	20.9	21.8	21.6	23.8
Val	13.9	14.6	14.3	15.6
Met	1.5	3.1	2.4	0.9
Ileu	11.9	11.6	11.8	12.4
Leu	15.7	14.9	15.6	16.7
Tyr	5.9	6.1	5.9	5.6
Phe	9.1	9.1	9.7	8.9
His	4.2	4.0	4.3	1.1
Lys	12.3	13.1	12.4	11.7
Arg	9.5	9.5	10.9	9.9

[a]   Moles of amino acid residue per subunit of PPO.
[b]   19.0 μM protein and 5 mM ascorbic acid in 0.1 M sodium phosphate buffer, pH 6.5, at 25°C for 22 h in a gyrotory water bath shaker.
[c]   19.0 μM protein and 0.8 mM cupric sulfate; other conditions as in "a".
[d]   19.0 μM protein, 5 mM ascorbic acid, and 0.8 mM cupric sulfate; other conditions as in "a".

glutamic acid/glutamine (from 26.9 to 31.2), glycine (from 19.9 to 23.3), and alanine (from 20.9 to 23.8) residues.

Subsequent studies showed that the amino acid compositions of bovine serum albumin, ovalbumin, and Kunitz soybean trypsin inhibitor were also modified by treatment with ascorbate-Cu^{2+} (Table 3). Only the amino acids listed showed a change. There was a small loss of histidine residues in bovine serum albumin and soybean trypsin inhibitor treated with ascorbate alone, but not for polyphenol oxidase and ovalbumin (Table 4). Cupric ion treatment alone caused a small loss of histidine residues in bovine serum albumin, ovalbumin, and soybean trypsin inhibitor, but not with polyphenol oxidase. With ascorbate-Cu^{2+}, there was a major loss of histidine in all four proteins: 3 of the 4 histidine residues of each subunit of polyphenol oxidase, 15 of the 16 histidine residues of bovine serum albumin, 5 of the 7 histidine residues of ovalbumin, and all 3 of the histidine residues of soybean trypsin inhibitor were lost. Ascorbate or Cu^{2+} alone caused little change in the methionine content, except for the Cu^{2+} treatment of soybean trypsin inhibitor (Table 5). Ascorbate-Cu^{2+} treatment caused more loss.

There was a correlation between the loss of histidine residues by ascorbate-Cu^{2+} treatment and the increase in aspartic acid residues (Table 6). In the case of ovalbumin, there was loss of five histidine residues and an increase of five aspartic acid residues. For the other three proteins, the stoichiometry was not as good (see below for further discussion). Cooper et al.[12] found a decrease in histidine and methionine and an increase in aspartic acid following incubation of some glycoproteins with Cu^{2+} and H_2O_2. Polyhistidine incubated under the same conditions lost 94.5% of the histidines, with formation of 80.3% aspartic acid residues.[12]

Size-exclusion chromatography (Figure 1) indicated that little fragmentation (other data indicate <5%) of polyphenol oxidase, bovine serum albumin, ovalbumin, and Kunitz soybean inhibitor occurred as a result of ascorbate-Cu^{2+} treatment. However, there was a marked change in the gel electrophoretic patterns, indicating that many new, more acidic residues were formed. The ascorbate-Cu^{2+}-treated proteins were found distributed throughout the gel (data not shown).

IV. DISCUSSION

Other researchers have also reported the effect of ascorbate-Cu^{2+} on proteins. Shinar et al.[25] reported that purified and membrane-bound acetylcholine esterase is rapidly inactivated by ascorbate-Cu^{2+}; they proposed that the mechanism of inactivation probably involved superoxide ions. Marx and Chevion[26] found that treatment of human and bovine serum albumin resulted in extensive molecular modification, indicated by decreased fluorescence and limited chain breaks. They proposed a mechanism involving ·OH radicals. Fibrinogen clots when treated with ascorbate-Cu^{2+}.[27] Clotting was prevented by citrate (a chelating agent) and by mannitol (a hydroxyl radical scavenger). Davison

TABLE 3
Changed Amino Acid Composition[a] of Proteins Incubated with Ascorbate-Cupric Sulfate

Amino acid	PPO		Ovalbumin		Soybean trypsin inhibitor[d]		Bovine serum albumin	
	Control[b]	Treated[c]	Control[b]	Treated[c]	Control[b]	Treated[c]	Control[b]	Treated[c]
His	4.2	1.1	7.2	1.8	2.9	0.0	15.7[e]	2.2
Met	1.5	0.9	11.7	9.4	3.1	1.4	3.1	2.4
Asx	27.4	33.9	35.3	40.0	27.1	27.5	51.2	60.6
Glx	26.9	31.2	44.3	47.2	25.2	24.6	73.8	80.9
Gly	19.9	23.3	21.6	25.5	16.0	16.5	15.3	19.3
Ala	20.9	23.8	31.2	32.8	7.5	7.8	39.3	45.6

[a]   Moles of amino acid residues per mole of protein, except for PPO where subunit mol wt of 28,000 is used.

[b]   Incubation conditions: 19.0 μM protein, 0.1 M sodium phosphate buffer, pH 6.5, at 25°C for 22 h in a gyrotory water bath shaker.

[c]   As in "a", except contained 5 mM ascorbic acid and 0.8 mM CuSO$_4$.

[d]   Kunitz soybean trypsin inhibitor of mol wt = 20,800.

[e]   Literature value for bovine serum albumin is 17 histidine residues per mole.

TABLE 4
Histidine Content of Proteins Incubated with Ascorbate, Cu^{2+}, or Ascorbate-Cu^{2+}

Conditions	Protein (moles histidine/mole protein)			
	Polyphenol oxidase[b]	Bovine serum albumin	Ovalbumin	Soybean trypsin inhibitor
Control[a]	4.2	15.7	7.2	2.9
+ Ascorbic acid (5 mM)	4.0	11.7	6.7	2.0
+ Cu^{2+} (0.8 mM)	4.3	11.7	6.5	2.1
+ Ascorbic acid-Cu^{2+} (5 mM/0.8 mM)	1.1	2.2	1.8	0.0

[a] Incubation conditions: 19 μM protein, 0.1 M sodium phosphate buffer, pH 6.5, at 25°C for 22 h.

[b] Moles of histidine per subunit weight of 28,000.

TABLE 5
Methionine Content of Protein Incubated with Ascorbate, Cu^{2+}, or Ascorbate-Cu^{2+}

Conditions	Protein (moles methionine/mole protein)			
	Polyphenol oxidase[b]	Bovine serum albumin	Ovalbumin	Soybean trypsin inhibitor
Control[a]	1.5	3.1	11.7	3.1
+ Ascorbic acid (5 mM)	3.1	2.4	12.5	1.8
+ Cu^{2+} (0.8 mM)	2.4	3.1	12.6	0.9
+ Ascorbic acid-Cu^{2+} (5 mM/0.8 mM)	0.9	2.4	9.4	1.4

[a] Incubation conditions: 19 μM protein, 0.1 M sodium phosphate buffer, pH 6.5, at 25°C for 22 h.

[b] Moles of methionine per subunit weight of 28,000.

et al.[28] found that catalase is reversibly inactivated by ascorbate, but when Cu^{2+} is added, the inactivation is irreversible. They proposed that H_2O_2 generated in the reaction is responsible for the inactivation. Harwood et al.[29] found that 3-hydroxy-3-methylglutaryl coenzyme A reductase activity in microsomes was markedly decreased by either ascorbate or dehydroascorbate in the range of 0.01 to 10 mM. They proposed that this may be responsible for the observed lowered cholesterol levels when large doses (500 to 4000 mg/d) of ascorbic acid are ingested. Glutamine synthetase was reported to be

TABLE 6
Aspartic Acid and Histidine Content[a] of Proteins Incubated with Ascorbate-Cu^{2+}

	Protein			
Conditions	Polyphenol oxidase[d]	Bovine serum albumin	Ovalbumin	Soybean trypsin inhibitor
Control[b]	27.4 (4.2)[e]	51.2 (15.7)	35.3 (7.2)	27.1 (2.9)
Ascorbic acid-Cu^{2+} (5 mM/0.8 mM)[c]	33.9 (1.1)	60.6 (2.2)	40.0 (1.8)	27.5 (0.0)

[a] Moles of amino acid per mole of protein, except for polyphenol oxidase. See "d".
[b] Incubation conditions: 19 μM protein, 0.1 M sodium phosphate buffer, pH 6.5, at 25°C for 22 h.
[c] As in "b", except for added ascorbate-Cu^{2+}.
[d] Residues per subunit weight of 28,000.
[e] Values in parentheses are for histidine.

inactivated by ascorbate-Fe^{2+} in the presence of O_2,[30,31] as a result of specific destruction of the one histidine residue at the active site of the enzyme.[32] Pyruvate kinase and creatinine kinase were also inactivated.[31] Insulin was denatured and became insoluble when treated with ascorbate-Cu^{2+}.[33]

A. MECHANISM OF ASCORBATE-Cu^{2+} ACTION ON PROTEINS

A mechanism for the action of ascorbate-Cu^{2+} on the four proteins examined in this study must account for the: (1) requirement for oxygen (see Table 1), (2) requirement for copper ions, (3) specificity for histidine residues of polyphenol oxidase, bovine serum albumin, ovalbumin, and soybean trypsin inhibitor, and (4) increase in aspartic acid, glutamic acid, glycine, and alanine residues.

The effect of Cu^{2+} and Fe^{2+} on the autoxidation of ascorbic acid has been studied since the 1930s.[34] Numerous workers have suggested detailed mechanisms based on the rate of dependence of the concentration of ascorbic acid, Cu^{2+} and H$^+$, and the products formed.[35] Khan and Martell,[36] based on extensive studies of the kinetics, suggested that the following reactions are necessary to account for the rate dependence and products formed. We shall replace Mn$^+$ with Cu^{2+} in their scheme.

$$AH_2 \overset{k_4}{\rightleftharpoons} AH^- + H^+ \tag{3}$$

$$AH^- + Cu^{2+} \overset{k_5}{\rightleftharpoons} Cu^{2+} \cdot AH^- \tag{4}$$

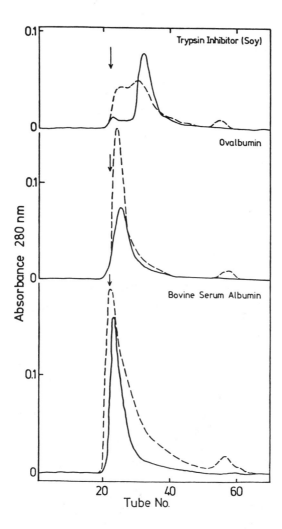

FIGURE 1. Elution pattern on Sephadex G-50 of ascorbic acid-cupric sulfate-treated and untreated proteins. The column (1.8 × 39.5 cm) was eluted as described in Section II.B. Ascorbic acid-cupric sulfate-treated (--) and untreated (—). Amounts of proteins placed on the column were: soybean trypsin inhibitor, 0.5 mg of untreated or treated; ovalbumin, 0.5 mg of untreated or treated; bovine serum albumin, 0.7 and 0.6 mg of untreated and treated, respectively. Arrows indicate void volume position.

$$\text{Cu}^{2+} \cdot \text{AH}^- + \text{O}_2 \underset{}{\overset{k_6}{\rightleftharpoons}} \text{Cu}^{2+} \cdot \text{AH}^- \cdot \text{O}_2 \tag{5}$$

$$\text{Cu}^{2+} \cdot \text{AH}^- \cdot \text{O}_2 \xrightarrow{k_7} \text{Cu}^{2+} \cdot \text{AH}^{\cdot} \cdot \text{O}_2 \tag{6}$$

$$\text{Cu}^{2+} \cdot \text{AH}^{\cdot} \cdot \text{O}_2 \xrightarrow{\text{fast}} \text{AH}^{\overline{\cdot}} + \text{Cu}^{2+} + \text{HO}_2^{\cdot} \tag{7}$$

$$\text{AH}^{\overline{\cdot}} + \text{Cu}^{2+} \xrightarrow{\text{fast}} \text{A} + \text{Cu}^+ \tag{8}$$

$$\text{AH}^{\overline{\cdot}} + \text{O}_2 \rightarrow \text{A} + \text{O}_2^{\overline{\cdot}} \tag{9}$$

$$\text{Cu}^+ + \text{HO}_2^{\cdot} + \text{H}^+ \xrightarrow{\text{fast}} \text{Cu}^{2+} + \text{H}_2\text{O}_2 \tag{10}$$

The mechanism proposed requires that ascorbic acid (AH_2) participate as the ascorbate anion (AH^-; Equation 3), that AH^- combine with Cu^{2+} and O_2 to form a ternary complex, $\text{Cu}^{2+} \cdot \text{AH}^- \cdot \text{O}_2$ (Equations 4 and 5), which then forms the radical $\text{Cu}^{2+} \cdot \text{AH}^{\cdot} \cdot \text{O}_2$ (Equation 6) that dissociates to give the semidehydroascorbate radical ($\text{AH}^{\overline{\cdot}}$), Cu^{2+}, and the peroxy radical $\text{HO}_2^{\cdot}$ (Equation 7). The semidehydroascorbate radical reacts with Cu^{2+} to form Cu^+ and dehydroascorbate (Equation 8) or with O_2 to form a superoxide radical ($\text{O}_2^{\overline{\cdot}}$) and dehydroascorbate (Equation 9). The Cu^+ is oxidized by the peroxy radical ($\text{HO}_2^{\cdot}$) to Cu^{2+} and H_2O_2 (Equation 10). Therefore, in this scheme, four strong oxidant species are formed: $\text{Cu}^{2+} \cdot \text{AH}^{\cdot} \cdot \text{O}_2$, $\text{AH}^{\overline{\cdot}}$, $\text{HO}_2^{\cdot}$, and $\text{O}_2^{\overline{\cdot}}$. Direct evidence for the formation of the ternary complex (catalyst · substrate · oxidant; $\text{Cu}^{2+} \cdot \text{AH}^- \cdot \text{O}_2$ above) has been presented by Skov and Vonderschmitt.[37]

More recently, Uchida and Kawakishi[14] presented a quite different scheme in which they included superoxide dismutase to remove the superoxide radical and catalase to remove the H_2O_2 formed (Scheme 1). It is clear that the final word on the mechanism of Cu^{2+}-catalyzed oxidation of ascorbate to dehydroascorbate has not been written.

The site-directed specificity toward histidine residues of proteins (and possibly disulfide bonds, thioether group of methionine, and indole of tryptophan) can best be explained by postulating that the Cu^{2+} involved in the reactions shown above (Equations 3 to 10) is bound to the imidazole group of histidine (or to the other residues listed) prior to reaction with ascorbate ion and O_2. Selective and tight complex formation of Cu^{2+} with free histidine, histidine peptides, and histidine residues of proteins is well documented. Doran et al.[38] reported the binding constants (reported as log K) of histidine and histidyl-histidine with Cu^{2+} to be 10.21 and 12.0, respectively, while that for gly-cylglycine was 6.52. Serum albumin has been reported to have several strong

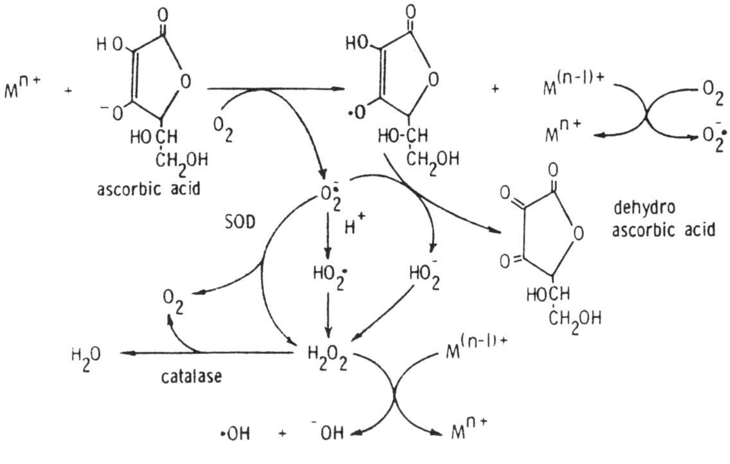

SCHEME 1. Proposed mechanism for metal ion-catalyzed oxidation of ascorbate. (From Uchida, K. and Kawakishi, S., *Agric. Biol. Chem.*, 50(2), 367, 1986. With permission.)

binding sites for metal ions. The N-terminal sequence Asp-Ala-His was reported to bind Cu^{2+} with $K_d = 6.61 \times 10^{-17}\ M^{-1}$,[39] with another tight binding site involving His-9 and His-18[39,40] and a third tight binding site involving Cys-34.[41] We have shown that histidyl-containing peptides protect polyphenol oxidase from inactivation in the presence of ascorbate-Cu^{2+} and that the histidine residue of these peptides becomes degraded in the process.[47] Fleming et al.[42] have shown that disulfides can complex directly with ascorbate, and tryptophan and tyrosine were reported to be lost in the presence of ascorbate[43] (no added metal ions). Martell[44] has presented evidence that metal chelates of ascorbate, including Cu^{2+}, are formed (see also Equation 4).

As shown in Table 3, histidine and methionine (to a lesser extent) are lost when polyphenol oxidase, bovine serum albumin, ovalbumin, and soybean trypsin inhibitor are incubated with ascorbate-Cu^{2+}. On the other hand, there is an increase in aspartic acid, glutamic acid, glycine, and alanine. Imanaga[45] treated histidine and histidine derivatives with ascorbate-Cu^{2+} and determined the ninhydrin-reactive products by paper chromatography. He reported the following products: glycine from imidazole carboxylic acid, aspartic acid (66% yield) from histidine, glutamic acid from imidazolepropionic acid and urocanic acid, alanine from methylimidazole, and serine from hydroxymethylimidazole. He found that addition of H_2O_2 to the reaction gave negligible degradation of hydroxymethylimidazole, compared to that produced by ascorbic acid. He suggested that this implicated the semidehydroascorbate radical as a participant in the reaction.

Aspartic acid is formed from histidine by photooxidation, by radiation, and by ascorbate-Cu^{2+}. The reaction of histidine is thought to involve fragmentation of the imidazole ring to give aspartic acid, urea, and other products (Equation 11). Indeed, we have shown[47] that the amount of urea formed is about 50% the amount of aspartic acid when bovine serum albumin is incubated with ascorbate-Cu^{2+}, as described in Table 2.

$$
\begin{array}{ccc}
\text{H}_2\text{N} & & \text{NH}_2 \\
| & & | \\
\text{HC} = \text{C–CH}_2\text{–C–COOH} \longrightarrow & \text{HOOC–CH}_2\text{–C–COOH} \\
\quad\quad\quad\quad | & & | \\
\quad\quad\quad\quad \text{H} & & \text{H} \\
\end{array}
$$

$$
\begin{array}{cc}
\text{HN} \quad \text{N} & \text{O} \\
\backslash \quad \slash\slash & \| \\
\text{C} & + \text{H}_2\text{N–C–NH}_2 \quad\quad (11) \\
| & \\
\text{H} & \\
\end{array}
$$

+ other products

We believe there is good indirect evidence that the reactive system leading to degradation of histidine residues in proteins is a quaternary complex between the imidazole group, Cu^{2+}, ascorbate, and O_2, i.e., $Im \cdot Cu^{2+} \cdot AH^- \cdot O_2$ (see Equation 5 for analogy), which then reacts to give $Im \cdot Cu^{2+} \cdot AH^{\cdot} \cdot O_2$. Whether the reactive species that degrades the histidine residue is $Im \cdot Cu^{2+} \cdot AH^{\cdot} \cdot O_2$ or $AH^{\cdot}$, $HO_2^{\cdot}$, $O_2^{\cdot}$, or H_2O_2 formed in proximity to the imidazole group (see Equations 7 to 10) is not known. We do know that H_2O_2 is produced in quite low amounts in the reaction (determination with peroxidase) and that H_2O_2 added to a control in much larger amounts than that formed had no effect on bovine serum albumin. Glucose and mannitol also did not decrease the rate of the reaction. Kim et al.[46] showed that 3-α-hydroxysteroid dehydrogenase was rapidly inactivated by superoxide radicals generated in the reaction by the aerobic xanthine oxidase reaction.

A great deal more research is needed to elucidate the mechanism of degradation of histidine and methionine residues of proteins on treatment with ascorbate-Cu^{2+}. This would be furthered by a better understanding of the mechanism of Cu^{2+}-catalyzed oxidation of ascorbate and the ascorbate-Cu^{2+}-catalyzed degradation of histidine peptides in model systems. A difficulty in elucidating the mechanism is determining the reactant species at the localized level as proposed in the quaternary complex. Bulk-added reactants and antagonist are not expected to cause an effect at the localized level.

ACKNOWLEDGMENT

The authors express their appreciation to BARD for support of this research. They also thank Virginia DuBowy and Pam Bains for typing the manuscript and checking the references.

REFERENCES

1. **Englard, S. and Seifter, S.**, *Annu. Rev. Nutr.*, 6, 365, 1986.
2. **Samuni, A., Chevion, M., and Czapski, G.**, *J. Biol. Chem.*, 256, 12632, 1981.
3. **Yamauchi, O. and Seki, H.**, *Chem. Lett.*, 1241, 1982.
4. **Florence, T. M., Stauber, J. L., and Mann, K. J.**, *J. Inorg. Chem.*, 24, 243, 1985.
5. **Iwata, M. and Kuzuhara, H.**, *Bull. Chem. Soc. Jpn.*, 58, 1609, 1985.
6. **Jones, G. J., Waite, D. W., and Smith, J. D.**, *Biochem. Biophys. Res. Commun.*, 128, 1031, 1985.
7. **Lipsky, P. E. and Ziff, M.**, *J. Immunol.*, 120, 1006, 1978.
8. **Lipsky, P. E. and Ziff, M.**, *J. Clin. Invest.*, 65, 1069, 1980.
9. **Starkebaum, G. and Root, R. K.**, *J. Immunol.*, 134, 3371, 1985.
10. **Munday, R.**, *J. Appl. Toxicol.*, 5, 69, 1985.
11. **Turok, K. and Joshi, S.**, *Eur. J. Biochem.*, 153, 155, 1985.
12. **Cooper, B., Creeth, J. M., and Donald, A. S. R.**, *Biochem. J.*, 228, 615, 1985.
13. **Mirvish, S. S.**, *Cancer*, 58 (Suppl.), 1842, 1986.
14. **Uchida, K. and Kawakishi, S.**, *Agric. Biol. Chem.*, 50, 367, 1986.
15. **Matsumura, G. and Pigman, W.**, *Arch. Biochem. Biophys.*, 110, 526, 1965.
16. **Murata, A., Kitagawa, K., and Saruno, R.**, *Agric. Biol. Chem. Jpn.*, 35, 294, 1971.
17. **Murata, A. and Kitagawa, K.**, *Agric. Biol. Chem. Jpn.*, 37, 1145, 1973.
18. **Stich, H. F., Karim, J., Koropatnick, J., and Lo, L.**, *Nature*, 260, 722, 1976.
19. **Chious, S.-H. and Ohtsu, N.**, *Proc. Natl. Sci. Counc. B. ROC*, 9, 275, 1985.
20. **Chiou, S.-H., Chang, W.-C., Jou, Y.-S., Chung, H.-M. M., and Lo, T.-B.**, *J. Biochem. (Tokyo)*, 98, 1723, 1985.
21. **Golan-Goldhirsh, A. and Whitaker, J. R.**, *J. Agric. Food Chem.*, 32, 1003, 1984.
22. **Golan-Goldhirsh, A. and Whitaker, J. R.**, in American Chemical Society, Division of Agricultural and Food Chemistry Meeting, Washington, D.C., 1984.
23. **Golan-Goldhirsh, A., Whitaker, J. R., Chen, A., and Osuga, D.**, in Federation of American Society of Experimental Biology Meeting, Anaheim, CA, 1985.
24. *Hoefer Scientific Instrument Manual,* Hoefer Scientific, San Francisco, CA, 1980.
25. **Shinar, E., Navok, T., and Chevion, M.**, *J. Biol. Chem.*, 258, 14778, 1983.
26. **Marx, G. and Chevion, M.**, *Biochem. J.*, 236, 397, 1985.
27. **Marx, G. and Chevion, M.**, *Thromb. Res.*, 40, 11, 1985.
28. **Davison, A. J., Kettle, A. J., and Fatur, D. J.**, *J. Biol. Chem.*, 261, 1193, 1986.
29. **Harwood, H. J., Jr., Greene, Y. J., and Stacpoole, P. W.**, *J. Biol. Chem.*, 261, 7127, 1986.
30. **Levine, R. L., Oliver, C. N., Fulks, R. M., and Stadtman, E. R.**, *Proc. Natl. Acad. Sci. U.S.A.*, 78, 2120, 1981.
31. **Levine, R. L.**, *J. Biol. Chem.*, 258, 11828, 1983.
32. **Levine, R. L.**, *J. Biol. Chem.*, 258, 11823, 1983.
33. **Inoue, H. and Hirobe, M.**, *Chem. Pharm. Bull.*, 34, 1075, 1986.
34. **Barron, E. S. G., DeMeio, R. H., and Klemperer, F.**, *J. Biol. Chem.*, 112, 625, 1936.
35. **Mushran, S. P. and Agrawal, M. C.**, *J. Sci. Ind. Res.*, 36, 274, 1977.
36. **Khan, M. M. T. and Martell, A. E.**, *J. Am. Chem. Soc.*, 89, 4176, 1967.
37. **Skov, K. A. and Vonderschmitt, D. J.**, *Bioinorg. Chem.*, 4, 199, 1975.
38. **Doran, M. A., Chaberek, S., and Martell, A. E.**, *J. Am. Chem. Soc.*, 86, 2129, 1964.
39. **Bradshaw, R. A., Shearer, W. T., and Gurd, F. R. N.**, *J. Biol. Chem.*, 243, 3817, 1968.
40. **Bradshaw, R. A. and Peters, T., Jr.**, *J. Biol. Chem.*, 244, 5582, 1969.
41. **Shaw, C. F., III, Schaeffer, N. A., Elder, R. C., Eidness, M. K., Trooster, J. M., and Calis, G. H. M.**, *J. Am. Chem. Soc.*, 106, 3511, 1984.
42. **Fleming, J. E., Bensch, K. G., Shreiber, J., and Lohmann, W.**, *Z. Naturforsch.*, 38c, 859, 1983.

43. **Torii, K. and Moriyama, T.,** *J. Biochem.,* 42, 193, 1955.
44. **Martell, A. E.,** in *Ascorbic Acid: Chemistry, Metabolism and Uses,* Seih, P. S. and Talbert, B. N., Eds., American Chemical Society, Washington, D.C., 1982, 153.
45. **Imanaga, Y.,** *J. Biochem.,* 42, 669, 1955.
46. **Kim, H.-S., Minard, P., Legoy, M.-D., and Thomas, D.,** *Biochem. J.,* 233, 493, 1986.
47. **Golan-Goldhirsh, A., Osuga, D. T., Chen, A. O., and Whitaker, J. R.,** unpublished observations.

Chapter 6

OXALYL THIOLESTERS, NOVEL INTERMEDIATES IN BIOLOGICAL SYSTEMS

Gordon A. Hamilton

TABLE OF CONTENTS

PRELUDE

An earlier version of this article was presented at a symposium held in May, 1988 to honor Myron L. Bender on the occasion of his retirement. At that time, the following comments were made by the current author prior to his scientific presentation. It seems appropriate to reproduce them here since the sentiments expressed still hold true, despite the subsequent passing of Myron.

> It is a real pleasure and honor to be able to participate in this symposium honoring Myron Bender on the occasion of his retirement. I know I feel, and probably it has been true for most of us who are former students of his, that the time spent in Myron's laboratory was not only one of the most exciting and intellectually stimulating times of our lives, but also one of the most influential. At that stage in one's life (in my case just after I had received my doctoral degree), when one is still trying to define where one is going, not only intellectually and scientifically, but also as a human being, the importance of having a mentor like Myron cannot be overstated. For, probably more than we realize, our whole approach to doing science is set during these formative years. It would be impossible to summarize all the characteristics that we, as former students, now have because we were associated with Myron, and probably we are not even aware of many of them. However, a few of his characteristics that stand out and were passed on to us include: his thoughtful approach to solving scientific problems; his infectious enthusiasm for science in general and especially for bioorganic chemistry; his constant striving for scientific truth, and doing the one experiment that will elucidate it in any particular instance; and his general drive to get research accomplished. These characteristics have all stood us in good stead in our careers, but, at least to me, the one quality of Myron that is his greatest legacy is that he showed us that one can do all the foregoing and still maintain basic human decency. Unlike what we might have learned from some others in this sometimes cutthroat field, Myron taught us that, in order to advance our scientific careers, one does not have to, for example, appropriate the ideas of others, or belittle and downgrade the work of other scientists. One can get ahead by just solid scientific accomplishment in one's own right. For showing us that, Myron, I certainly, and I think most of us, thank you; it is one of the main reasons we are here today to participate in this symposium to honor you.

It is also one of the main reasons this article and this volume are being dedicated to the memory of Myron L. Bender.

I. INTRODUCTION

Oxalyl thiolesters ($RSCOCOO^-$) are a newly discovered class of mammalian metabolites that appear to play an important role in controlling animal metabolism, and that may also participate in the intracellular messenger system for some hormones, especially insulin and growth factors. In this article is reviewed the research performed in our laboratory that has led to the discovery of these metabolites, clarified their chemical and biochemical reactivity, and led to the conclusion that they may be especially important metabolites.

II. INITIAL INDICATIONS FOR OXALYL THIOLESTERS (OTEs) IN ANIMALS

Several years ago, we obtained evidence that the actual physiological substrates for a group of mammalian peroxisomal oxidases (D-amino acid oxidase, D-aspartate oxidase, and L-hydroxy acid oxidase) are not the stable substrates usually associated with these enzymes, but are adducts of glyoxylate with various thiols and aminothiols.[1-9] For the enzyme L-hydroxy acid oxidase, the best substrates turn out to be glyoxylate thiohemiacetals, and the product of the enzymic reaction (Equation 1) is an oxalyl thiolester (OTE). Following a detailed investigation of the specificity of this reaction,[1,4-8] we became convinced that the reaction had to be occurring physiologically; in some cases, the thiohemiacetals are one to two orders of magnitude better as substrates than previously studied simple hydroxy acid substrates. This was the first indication that OTEs might be present in animals. Subsequently, the possibility that OTEs might also be formed from the suspected physiological products of the D-amino acid and D-aspartate oxidase-catalyzed reactions,[10] as well as from ascorbic acid,[11] were also considered.

$$
RSH + \underset{\text{glyoxylate}}{\overset{\displaystyle \underset{|}{\overset{CH=O}{|}}}{COO^-}} \xrightarrow[\text{enzymic}]{\text{non}} \underset{\text{thiohemiacetal}}{\overset{\displaystyle \overset{OH}{|}}{RS-CH-COO^-}} \xrightarrow[\underset{O_2 \quad H_2O_2}{}]{\text{enzyme}} \underset{\text{OTE}}{\overset{\displaystyle \overset{O}{\|}}{RS-C-COO^-}} \quad (1)
$$

III. EARLY INDICATIONS FOR A POSSIBLE PHYSIOLOGICAL ROLE FOR OTEs

Having concluded that OTEs are probably produced in animals, the next question that arises is: what are the physiological functions of these compounds? In an attempt to answer this question, the known physiological effects of a large number of compounds were correlated[1,8,12,13] with their ability to inhibit the oxidase (although the initial correlations were made specifically for the D-amino acid oxidase reaction, very similar correlations hold for the hydroxy acid oxidase reaction as well, since the two enzymes are inhibited by similar compounds[14]). As a result, it was concluded that OTEs are probably involved in controlling metabolism, and are possibly functioning as part of an intracellular messenger system for some hormones, especially insulin. The correlation with insulin effects is particularly noteworthy; compounds that inhibit the oxidase reactions have insulin-like effects physiologically, while those that speed up these reactions have antiinsulin effects. This, therefore, indicates that if OTEs are part of an intracellular messenger system for insulin, they are negative messengers, i.e., insulin should decrease their concentration.

Consequently, any such putative effector should have the opposite effect on intracellular enzymes and metabolism that insulin does. Since insulin and animal growth factors have very similar metabolic effects, the implication of the foregoing correlations is that OTEs may also play a role in the control of cell growth and proliferation. If so, then the correlations predict, as before, that growth factors should cause a decrease in intracellular concentrations of OTEs and that their concentrations should be higher in nonproliferating cells.

IV. CHEMICAL REACTIVITY OF OTEs

Because of the evidence that OTEs may be intracellular hormonal mediators and may be modifying the activities of various enzymes, an extensive investigation of their chemical reactivities was carried out[1,8,15] in order to better understand how they might be accomplishing that. Under physiological conditions, OTEs are stable to hydrolysis and also react too slowly with typical physiological nitrogen nucleophiles for the nonenzymic reaction to have any metabolic significance. In contrast, OTEs react surprisingly rapidly with thiols (Equation 2), with the halftime for the nonenzymic oxalyl transfer being only seconds to minutes (depending on the thiol concentration) at pH 7.4 and 37°C. This means that if OTEs are present physiologically, then the oxalyl group will be exchanging, by a nonenzymic reaction, within seconds to minutes with virtually every thiol in the cell. Since glutathione (GSH) and CoASH are the main small-molecule thiols in the cell, GS-Ox and CoAS-Ox will be the main small-molecule OTEs. However, free thiol groups are components of many enzymes (ESH) and proteins, so one suspects that oxalyl derivatives of enzymes (ES-Ox) will also be formed (see Section V) if OTEs are present *in vivo*.

$$\text{RS–C–COO}^- + \text{R'SH} \rightleftarrows \text{R'S–C–COO}^- + \text{RSH} \qquad (2)$$
$$\overset{\|}{\underset{O}{}} \qquad\qquad\qquad \overset{\|}{\underset{O}{}}$$

The reaction of OTEs with 2-aminoethanethiols at pH 7.4 leads to the formation of stable *N*-oxalyl derivatives by the mechanism of Equation 3; the initial oxalyl transfer to the thiol group occurs as readily as with other thiols, but this is immediately followed by a very rapid (halftime about 0.1 s at pH 7.4 and 37°C) *irreversible* intramolecular S to N migration. Thus, to the extent that the reaction of Equation 3 occurs, either *in vivo* or in experimental system, it leads to the removal of OTEs. The rate of the initial step in Equation 3 depends on the concentration of the 2-aminoethanethiol as well as on the temperature and pH. The main 2-aminoethanethiol expected *in vivo* is L-cysteine (Z = COO$^-$) and, at its suspected physiological concentration of 10 to 100 μM,[16] the halftime for the nonenzymic reaction of Equation 3 will be on the order of 10 min to a couple of hours at pH 7.4 and 37°C. Another known metabolite containing the 2-aminoethanethiol structure is L-cysteinyl-glycine ($Z = CONHCH_2COO^-$), but since its intracellular concentration is not known, one cannot estimate how important the nonenzymic reaction of Equation 3 with this compound will be *in vivo*. Finally, cysteamine ($Z = H$) also has the 2-aminoethanethiol structure, but its concentration is believed to be so low[17] *in vivo* that the nonenzymic reaction of Equation 3 with cysteamine in cells is expected to be almost negligible.

$$
\underset{\text{OTE}}{\overset{\overset{\displaystyle O}{\overset{\displaystyle \|}{}}}{RS-C-COO^-}} \; + \; \underset{\substack{\text{2-amino-}\\\text{ethanethiol}}}{\overset{\overset{\displaystyle CH_2-SH}{\overset{\displaystyle |}{}}}{CHZ-NH_2}} \; \rightarrow \; \overset{\overset{\displaystyle O}{\overset{\displaystyle \|}{}}}{CH_2-S-C-COO^-} \; + \; RSH
$$
$$
\underset{CHZ-NH_2}{|}
$$

$$
\xrightarrow{\text{fast}} \; \underset{\underset{\displaystyle O}{\overset{\displaystyle \|}{}}}{CHZ-NH-C-COO^-}
$$
$$
\overset{\overset{\displaystyle CH_2-SH}{\overset{\displaystyle |}{}}}{}
$$

(3)

N-oxalyl derivative

V. EFFECTS OF OTEs ON THE ACTIVITIES OF ENZYMES

As indicated above, one suspects that enzymes (ESH) with reactive thiol groups could react with OTEs to produce ES-Ox derivatives. Since other covalent modifications of such enzymic thiols are known frequently to alter the enzymic activity, one suspects that ES-Ox and ESH might have different activities. This offers one possible mechanism by which OTEs could affect enzymic activities if they are intracellular mediators for some hormones. However, there are other mechanisms as well. To determine whether OTEs have any effect on important enzymes and what the mechanism might be, we have investigated their effects on several different enzymes, especially those known to have reactive thiol groups and thought to be modified by insulin and/or growth factors. The results to date, as illustrated by the three examples given below, strongly imply that OTEs could indeed be important intracellular mediators for such hormones.

One group of enzymes that is thought to be affected when cells are stimulated with insulin are various protein phosphatases, and insulin is believed to cause an increase in their catalytic activities.[18] Since the previously mentioned correlations indicate that the oxidase products should have the opposite effect of insulin, OTEs should therefore inhibit the phosphatases if they are the responsible intracellular reagents. That is indeed observed with at least one of the protein phosphatases, the catalytic subunit of phosphorylase phosphatase.[19] Some noteworthy results obtained on the inhibition of this phosphatase by OTEs are the following: (1) the enzyme is inhibited by a number of OTEs, but it shows high specificity for *S*-oxalylglutathione (GS-Ox), which inhibits the enzyme more than an order of magnitude better than any other OTE tested, (2) inhibition is detectable with 25 μM or less concentrations of GS-Ox, and (3) the inhibition is time dependent and partially reversed by thiols, thus suggesting that at least part of the inhibition is due to oxalylation of an enzymic thiol (formation of ES-Ox) by the reaction of Equation 2. The fact that GS-Ox is such a good inhibitor lends support to the possibility that it could be an important effector in animal cells. Glutathione is the most abundant small-molecule thiol in the cell, so at equilibrium, GS-Ox would be the most abundant OTE. Furthermore, the high specificity suggests that the enzyme has evolved a specific binding site for GS-Ox, which in turn implies that it is a normal effector of the enzyme *in vivo*.

Another enzyme system that the evidence indicates insulin modifies is the pyruvate dehydrogenase complex (PDC) of mitochondria, and again the catalytic activity of this complex is increased when cells are stimulated with insulin.[20] Thus, as before, one predicts that OTEs should decrease the PDC activity if they are the responsible intracellular mediators. The PDC system is complex because not only are there three proteins involved in the overall reaction, but the system is also affected by phosphorylation and dephosphorylation catalyzed by a specific PDC kinase and a PDC-phosphate phosphatase. In recent experiments,[21] it has been found that OTEs do inhibit PDC activity. Interestingly, GS-Ox at low micromolar concentrations has very little effect, but CoAS-Ox is an effective inhibitor of the overall reaction (inhibition is easily detected at 10 μM concentrations of CoAS-Ox). From a detailed investigation of the mechanism of the inhibition, it has been clearly shown that the inhibition is due to the transfer of oxalyl groups from CoAS-Ox to the dihydrolipoyl residues of the PDC. When PDC is oxalylated, the rates of the reactions catalyzed by PDC kinase and PDC-phosphate phosphatase are also affected. The finding that PDC is inhibited specifically by CoAS-Ox adds further credence to the suggestion that OTEs may be normal intracellular enzyme effectors.

The third enzyme for which some data concerning its inhibition by OTEs[8,22] is available is succinyl-CoA transferase, which catalyzes the reversible transfer of the CoA group from succinyl-CoA to acetoacetate to give acetoacetyl-CoA and succinate. This enzyme is important in the utilization of acetoacetate in tissues other than liver, so its inhibition could partially explain the increased amounts of acetoacetate in the blood of diabetics or those with insulin insufficiency. Thus, our correlations again predict that OTEs should inhibit the enzyme if they are the responsible intracellular messenger. As with the PDC complex, we have found (using the pig heart enzyme) that CoAS-Ox is an effective inhibitor, while GS-Ox and other OTEs inhibit very little even at high concentrations. Interestingly in this case, the inhibition by CoAS-Ox appears to be due to simple binding to the enzyme, because it is not time dependent and is completely reversed when CoAS-Ox is removed by gel filtration chromatography. From double-reciprocal plots of data obtained at differing succinyl-CoA and CoAS-Ox concentrations, it was found that the inhibition is noncompetitive vs. succinyl-CoA, and K_{is} (at 25°C and pH 8.1) is 5 μM. The high specificity of the enzyme for CoAS-Ox vs. GS-Ox is perhaps not too surprising since its substrate is an acyl CoA, but CoAS-Ox binds even better than the normal substrate since the K_m for succinyl-CoA under the same conditions is 0.16 mM. In general, the high specificity for the oxalyl derivative of a physiological thiol, and the fact that inhibition is detectable at such low concentrations, argue strongly for a role for this compound in controlling the activity of the transferase *in vivo*.

It is of some interest that the last two enzymes that are inhibited specifically by CoAS-Ox are mitochondrial enzymes where the CoASH concentration[23] is known to be high (greater than 2 mM). Thus, higher concentrations of CoAS-Ox are expected in mitochondria than in the cytosol, where CoASH concentrations are much lower.

VI. IDENTIFICATION AND QUANTITATION OF OTEs AND *N*-OXALYLCYSTEINE AS ANIMAL METABOLITES

Since, prior to the present work, OTEs were not known animal metabolites, an HPLC assay to identify and quantitate them in biological samples was developed.[24,25] In this work, no attempt to quantitate individual OTEs was made because they exchange oxalyl groups so readily (Equation 2); rather, OTEs were analyzed for as a group. To do this, advantage was taken of the reaction of 2-aminoethanethiols with OTEs to give *N*-oxalyl derivatives by the mechanism of Equation 3. These *N*-oxalyl derivatives are stable and can be identified and quantitated by HPLC after derivatizing the thiol group with a fluorescent label. By this technique, down to 0.2 μM concentrations of OTEs in tissues can be quantitated.

A result obtained, not unexpectedly because of the data given earlier, is that *N*-oxalylcysteine is present in various tissue homogenates that are not treated with any external 2-aminoethanethiol, i.e., after the tissue is merely homogenized and immediately derivatized with the fluorescent label. The amount in various rat tissues,[24] such as kidney, liver, brain, heart, muscle, and fat, is on average approximately 10 nmol/g wet weight, but the values range from 1 to 18 nmol/g with different animals and tissues. Because of the considerations given earlier, the finding of *N*-oxalylcysteine in untreated tissue is an initial strong indication that OTEs are present *in vivo*.

If, prior to derivatization with the fluorescent label, homogenates are heated with excess added 2-aminoethanethiol (usually cysteamine is used for these analytical purposes) under conditions known to convert all OTEs to N-oxalyl derivatives by the reactions of Equation 3, then the amount of N-oxalylcys-teamine is increased dramatically (N-oxalylcysteamine is either negligible or just barely detectable in homogenates that have not been treated with added cysteamine). Such results indicate that OTEs are present in the homogenates, and this conclusion has been amply substantiated by subsequent control experiments.[25] The concentrations of OTEs in tissues vary somewhat from animal to animal, but in mature rat tissue the average amount is approximately 45 to 48 nmol/g wet weight in kidney and brain, about 23 nmol/g in liver and heart, and 4 to 8 nmol/g in epididymal fat and leg muscle. Thus, OTEs are present in animal tissues in surprisingly large quantities; disregarding possible compartmentation, the results mean that the aqueous portion of brain and kidney has a concentration of OTEs in excess of 60 μM, and liver and heart greater than 25 μM. These are concentrations that have been shown to affect the activities of various enzymes (see Section V), so it seems almost certain that OTEs will be affecting enzyme activities *in vivo*.

Since obtaining the foregoing results, many different biological samples have been analyzed, and in *all* mammalian tissues examined, both OTEs and N-oxalylcysteine have been detected. Furthermore, it has recently been found[26] that N-oxalylcysteine is also present in human urine (approximately 3 to 6 μM). The ubiquity of these new metabolites is another indication that they may have an important physiological function.

VII. PROBABLE ROLE OF OTEs IN INHIBITING CELL PROLIFERATION

Given the many indications discussed previously that OTEs may be important intracellular mediators for growth factors and cell growth stimulants, recently the levels of OTEs in bovine lymph node lymphocytes, when the lymphocytes are either resting or stimulated to proliferate, have been compared. In work that has been published,[27] it was found that treatment of lymphocytes with plant lectins (concanavalin A and phytohemagglutinin) and a phorbol ester tumor promoter, either alone or in combination, leads to a decrease in the OTE concentration, and the amount of the decrease correlates with the mitogenic response of the lymphocytes to the particular stimulant or stimulants. Thus, the greatest decrease in OTE concentration occurs with stimulation by either concanavalin A or a combination of phytohemagglutinin and phorbol ester, conditions that are the most mitogenic for these cells. The decrease in OTE concentration occurs within 1 h after stimulation and the levels remain low for up to 48 h. In later work,[26] it was found that similar effects are seen when lymphocytes are stimulated with a combination of wheat germ agglutinin and phorbol ester. Stimulation with either alone leads to only small or negligible decreases in OTE concentrations, but when the lymphocytes are treated with both together, a large decrease in OTE concentration is seen. Wheat germ agglutinin alone is not mitogenic for these cells and phorbol ester is only weakly mitogenic, but the two together are strongly mitogenic.

The foregoing results strongly imply that OTEs do have a role in the proliferation process and that they may be cell proliferation inhibitors since their concentrations decrease, as predicted by the earlier correlations, when cells are stimulated to proliferate. Other recent results are also in accord with this conclusion. Thus, it has been found[26] that addition of OTEs to the culture medium inhibits the proliferation response of lymphocytes to concanavalin A. Furthermore, in some preliminary experiments that need to be repeated, it was observed that the concentration of OTEs in a transformed (with SV40 virus) cell line of Swiss 3T3 fibroblasts is only approximately one fourth the concentration in resting normal 3T3 fibroblasts.[28]

One criticism of the above conclusion that OTEs may be cell proliferation inhibitors is that the measured decreases in OTE concentrations, when cells are stimulated to proliferate, may just be coincidental. However, some results obtained[29] using the enzyme γ-glutamyltransferase (frequently called transpeptidase), when considered in conjunction with literature information, make it even less likely that the decreases in OTE concentration on cell stimulation are just coincidental. The transferase, which is a plasma membrane-bound enzyme, catalyzes the first step in the metabolism of glutathione and glutathione conjugates,[30] namely, the transfer of the γ-glutamyl residue of such compounds to an amino acid, peptide, or water. The other product is cysteinylglycine, if glutathione itself is the reactant, or the S-substituted cysteinylglycine, if a glutathione conjugate is the reactant. From the known specificity of the enzyme, one could readily predict that GS-Ox should be a substrate, so the actual experiments showing that it is a good substrate (it has a K_m and V_{max} comparable to other good substrates, such as γ-glutamyl p-nitroanilide) are in some respects trivial. However, the implications of this result are considerable. When GS-Ox is the substrate, the product of the reaction is S-oxalylcysteinylglycine, which, being a β-amino-S-oxalyl derivative, will immediately rearrange (last part of Equation 3, Z = $CONHCH_2COO^-$) to N-oxalylcysteinylglycine. Indeed, the enzymic reaction can be monitored by following the disappearance of the thiolester absorbance at 260 nm. Consequently, to the extent that the γ-glutamyltransferase reaction on GS-Ox occurs in vivo, it effectively will decrease the cellular concentration of OTEs.

What makes the results with γ-glutamyltransferase of so much interest is the fact that the catalytic activity of this enzyme is increased severalfold in many tumors.[31] Although this increase in activity has been used as a marker for neoplasia, a suitable explanation for its dramatic increase in cancer tissue had not previously been articulated. We have proposed[29] that a major function of the transferase in normal cells is to control the level of OTEs in vivo, and that those cancers with elevated levels of transferase have found this way to lower the OTE concentration, thus relieving the inhibition of cell growth that we believe these compounds are responsible for. Although it is somewhat of a circular argument, the finding of the transferase-catalyzed reaction of GS-Ox thus makes it more likely that the decrease in OTE concentration, when cells are stimulated to proliferate, is not just a coincidence.

VIII. PROBABLE ROLE OF OTEs AS PART OF THE INTRACELLULAR MESSENGER SYSTEM FOR INSULIN

One of the well-characterized systems known to be sensitive to insulin is the rat epididymal fat cell system in which insulin stimulates the uptake of glucose and glucose derivatives.[32] Consequently, this system has recently been used[33] in an attempt to obtain evidence that OTEs may be involved in the response of the adipocytes to insulin. The initial plan was to measure the concentration of OTEs in the unstimulated and stimulated adipocytes, but unfortunately these cells contain so much fat that a large enough sample could not be obtained to analyze and detect OTEs by the HPLC method (the OTEs, being charged, will be present in the aqueous part of the cell). However, some evidence was obtained using this system that OTEs may indeed participate in the cellular response to insulin stimulation. The experiments involved comparing the rate of glucose uptake into resting cells, cells stimulated by insulin, and cells that had been incubated for a few minutes with approximately 1 mM 2-aminoethanethiol (cysteamine or cysteine). These last conditions should remove OTEs by the reaction of Equation 3, so if insulin acts by lowering the OTE concentration, then, following incubation with cysteamine or cysteine, the rate of glucose uptake should be increased as it is with insulin. That is what was observed. In fact, cells incubated with the 2-aminoethanethiols tend to take up glucose even faster than insulin-treated cells. This is not an effect just of thiols, because treatment of these cells with other thiols that are not capable of removing OTEs by the reaction of Equation 3 is known to not mimic the insulin stimulation of glucose uptake with these cells.

IX. CURRENT HYPOTHESIS ON HOW OTEs PARTICIPATE IN SIGNAL TRANSDUCTION

The results obtained to this point not only suggest strongly that OTEs are involved in signal transduction, but they also are beginning to indicate in outline how they may be involved. It is clear that changes in OTE concentrations do not occur as the very primary event when cells are treated with hormones and stimulants; for many of the hormones and stimulants, the primary events are known and they do not involve OTE metabolism. From the time course (usually within minutes) of the changes seen, however, it appears that the OTEs participate at a relatively early stage of the signal transduction sequence. Also, since the OTEs have been found to directly affect the catalytic activities of enzymes, it seems likely that they participate as ultimate effectors of the stimulant signal. Our current hypothesis is that insulin and the other cell stimulants either inhibit the rate of biosynthesis of the OTEs or increase the rate of their catabolism, possibly by causing a phosphorylation or dephosphorylation of one or more key enzymes in these processes. It is further proposed that interaction of the OTEs with enzymes and control proteins leads to the subsequent changes in cellular metabolism, including the activation or inhibition of certain genetic elements that cause changes in the rate of specific protein synthesis.

X. GENERAL COMMENTS

If one peruses the literature on the treatment of cultured mammalian cells, one is struck by how frequently it is recommended that the cell medium contain a 2-amino- or 2-hydroxy-substituted thiol, such as cysteine or mercaptoethanol. This is especially so if the cultured cells are being stimulated to grow and proliferate. In light of the current results, the requirement for such a substituted thiol is now apparent. Presumably, the removal of OTEs by the substituted thiol by the mechanism of Equation 3 (2-mercaptoethanol should react in a manner very similar to the 2-aminoethanethiols) is required in order for the cells to remain viable under the culture conditions and proliferate. It is clear, however, that the removal of OTEs is not the only requirement for cell proliferation, since cells incubated in a medium containing cysteine or mercaptoethanol do not proliferate until treated with other factors. Consequently, it appears that lowering OTE concentrations is a *necessary* but *not sufficient* requirement for cell proliferation.

Another process affecting mammalian systems in which 2-aminoethanethiols play a unique role is in the protection of animals against the lethal effects of ionizing radiation.[34] Although thiols in general have some protective effect, presumably because they minimize oxidative damage to the cells when irradiated, 2-aminoethanethiols (especially cysteamine) are much more effective as protectors than one would predict on the basis of them being thiols alone. In light of the current results, it now seems probable that removing OTEs may be important to cell and animal survivability following radiation. According to the aforementioned correlations and results, the removal of OTEs should lead to a more anaerobic metabolism (insulin-like effects) and an increase in the ability of cells to divide and reproduce themselves. Perhaps these are important to survivability following irradiation.

XI. SUMMARY

As indicated in the foregoing, there is now evidence that OTEs are present in animals and that they affect the activities of several important enzymes at measured physiological concentrations. Furthermore, evidence has been obtained that OTEs are cell proliferation inhibitors and that they may be part of the intracellular messenger system for insulin. Therefore, it appears that OTEs may be very important metabolic regulators.

ACKNOWLEDGMENTS

The author would like to acknowledge and thank his many co-workers who contributed so much to the development of this research; their names can be found in the reference list. The collaboration of Dr. Andrea M. Mastro in the experiments using animal cells in culture is also greatly appreciated. This research was supported in part by a grant from Eastman Kodak Company (Rochester, NY) and by research grants (DK 13448 and DK 38632) from the National Institute of Diabetes and Digestive and Kidney Diseases, Public Health Service (Bethesda, MD).

REFERENCES

1. **Hamilton, G. A.**, *Adv. Enzymol. Relat. Areas Mol. Biol.*, 57, 85, 1985.
2. **Hamilton, G. A., Buckthal, D. J., Mortensen, R. M., and Zerby, K. M.**, *Proc. Natl. Acad. Sci. U.S.A.*, 76, 2625, 1979.
3. **Naber, N., Venkatesan, P. P., and Hamilton, G. A.**, *Biochem. Biophys. Res. Commun.*, 107, 374, 1982.
4. **Brush, E. J. and Hamilton, G. A.**, *Biochem. Biophys. Res. Commun.*, 103, 1194, 1981.
5. **Hamilton, G. A. and Brush, E. J.**, *Dev. Biochem. Flavins Flavoproteins*, 21, 244, 1982.
6. **Brush, E. J. and Hamilton, G. A.**, *Ann. N.Y. Acad. Sci.*, 386, 422, 1982.
7. **Gunshore, S., Brush, E. J., and Hamilton, G. A.**, *Bioorg. Chem.*, 13, 1, 1985.
8. **Hamilton, G. A., Afeefy, H. Y., Al-Arab, M. M., Brush, E. J., Buckthal, D. J., Burns, C. L., Harris, R. K., Ibrahim, D. A., Kiselica, S. G., Law, W. A., Ryall, R. P., Skorczynski, S. S., and Venkatesan, P. P.**, in *Peroxisomes in Biology and Medicine*, Fahimi, H. D. and Sies, H., Eds., Springer-Verlag, Heidelberg, 1987, 223.
9. **Burns, C. L., Main, D. E., Buckthal, D. J., and Hamilton, G. A.**, *Biochem. Biophys. Res. Commun.*, 125, 1039, 1984.
10. **Afeefy, H. Y. and Hamilton, G. A.**, *Bioorg. Chem.*, 15, 262, 1987.
11. **Al-Arab, M. M. and Hamilton, G. A.**, *Bioorg. Chem.*, 15, 81, 1987.
12. **Hamilton, G. A. and Buckthal, D. J.**, *Bioorg. Chem.*, 11, 350, 1982.
13. **Hamilton, G. A., Buckthal, D. J., and Kalinyak, J.**, in *Oxidases and Related Redox Systems*, King, T. E., Mason, H. S., and Morrison, M., Eds., Pergamon Press, New York, 1982, 447.
14. **Brush, E. J.**, Ph.D. thesis, Pennsylvania State University, University Park, 1984.
15. **Law, W. A. and Hamilton, G. A.**, *Bioorg. Chem.*, 14, 378, 1986.
16. **Cooper, A. J. L.**, *Annu. Rev. Biochem.*, 52, 187, 1983.
17. **Ziegler, D. M.**, *Annu. Rev. Biochem.*, 54, 305, 1985.
18. **Cohen, P.**, *Annu. Rev. Biochem.*, 58, 453, 1989.
19. **Gunshore, S. and Hamilton, G. A.**, *Biochem. Biophys. Res. Commun.*, 134, 93, 1986.

20. **Ohlen, J., Siess, E. A., Loffler, G., and Wieland, O. H.**, *Diabetologia,* 14, 135, 1978.
21. **Liu, Y.**, Ph.D. thesis, Pennsylvania State University, University Park, 1991.
22. **Churley, M., Ibrahim, D. A., and Hamilton, G. A.**, *Fed. Proc.,* 46 (Abstr. 152), 1951, 1987.
23. **Robishaw, J. D. and Neely, J. R.**, *Am. J. Physiol.,* 248, E1, 1985.
24. **Skorczynski, S. S. and Hamilton, G. A.**, *Biochem. Biophys. Res. Commun.,* 141, 1051, 1986.
25. **Skorczynski, S. S., Yang, C.-S., and Hamilton, G. A.**, *Anal. Biochem.,* 192, 403, 1991.
26. **Yang, C.-S.**, Ph.D. thesis, Pennsylvania State University, University Park, 1991.
27. **Skorczynski, S. S., Mastro, A. M., and Hamilton, G. A.**, *FASEB J.,* 3, 2415, 1989.
28. **Skorczynski, S. S.**, Ph.D. thesis, Pennsylvania State University, University Park, 1989.
29. **Hamilton, G. A., Buckthal, D. J., Kantorczyk, N. J., and Skorczynski, S. S.**, *Biochem. Biophys. Res. Commun.,* 150, 828, 1988.
30. **Meister, A. and Anderson, M. E.**, *Annu. Rev. Biochem.,* 52, 711, 1983.
31. **Hanigan, M. H. and Pitot, H. C.**, *Carcinogenesis,* 6, 165, 1985.
32. **Simpson, I. A. and Cushman, S. W.**, *Biochem. Actions Horm.,* 13, 1, 1986.
33. **Libby, M. C. and Hamilton, G. A.**, unpublished results.
34. **Giambaressi, L. and Jacobs, A. J.**, in *Military Radiobiology,* Conklin, J. J. and Walker, R. I., Eds., Academic Press, New York, 1987, 265.

Chapter 7

CYCLODEXTRIN-INDUCED REGIOSELECTIVE CLEAVAGES OF RIBONUCLEIC ACIDS

Makoto Komiyama

TABLE OF CONTENTS

I. INTRODUCTION

Ribonuclease cleaves ribonucleic acids selectively to the fragments having the terminal phosphates at the 3'-positions (III-p in Figure 1).[1,2] Here, the 2',3'-cyclic monophosphate of the terminal ribonucleotide (I-p) is formed as the intermediate, and its P-O(2') bond is regioselectively cleaved. Attempts to mimic this enzyme were made.[2-6] However, regioselective cleavages of the P-O(2') bonds of 2',3'-cyclic monophosphates of ribonucleotides by artificial systems have not been successful yet.[6]

This paper reports the regioselective cleavages of the 2',3'-cyclic monophosphates of ribonucleosides, cytidine, uridine, adenosine, and guanosine (Ia-d), catalyzed by cyclodextrins (CyDs) (Equation 1).[7] Marked dependence of the regiospecific catalysis on the kind of CyD is shown. Furthermore, the CyD-induced regioselective cleavages of 3'-5'-linked polymers of ribonucleotides are presented.

(1)

B = cytosine, uracil, adenine, or guanine

II. EXPERIMENTAL

Cleavages were followed by periodic analyses with HPLC. All the reactions satisfactorily followed the pseudo first-order kinetics, and the selectivities were independent from the conversions.

III. RESULTS

A. REGIOSELECTIVE CLEAVAGE OF I INDUCED BY CyD[8-13]

As shown in Table 1, the P-O(2') bond of Ia is selectively cleaved in the presence of α-CyD, giving IIIa in high selectivity. The selectivity as well as the rate of cleavage asymptotically increases with an increase in the concentration of α-CyD. The selectivity attains 98% at a 0.05 M concentration of α-CyD.

In the absence of α-CyD, however, the cleavage of the P-O(3') bond takes place more efficiently than that of the P-O(2') bond, and the selectivity for IIIa is only 47%.

According to a kinetic analysis of the data, the rate constant for the P-O(2') cleavage of Ia in the α-CyD-Ia complex is 13.8 times as large as that of free Ia. On the other hand, the P-O(3') cleavage of Ia is totally suppressed in the α-CyD complex.

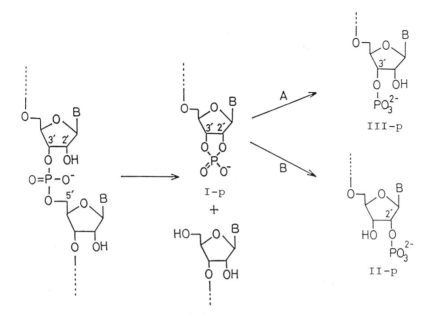

FIGURE 1. Scheme for the cleavages of ribonucleic acids.

TABLE 1
Selectivities and Rate Constants for the Cleavage of Ia in the Presence and Absence of CyDs[a]

CyD	Concentration (10^{-2} M)	Rate constant (10^{-4} min^{-1})	Selectivity[b] (%)
α-CyD	1.0	3.4	81
	5.0	7.0	98
β-CyD	1.0	1.9	45
γ-CyD	1.0	1.4	47
Hexa-2,6-dimethyl-α-CyD	1.0	2.0	47
Hepta-2,6-dimethyl-β-CyD	1.00	1.7	47
None	—	1.7	47

[a] pH 11.08, 20°C.
[b] IIIa/(IIa + IIIa).

Neither β-CyD nor γ-CyD shows a measurable increase of the selectivity. The catalyses by hexa-2,6-dimethyl-α-CyD and hepta-2,6-dimethyl-β-CyD are negligible.

The P-O(2′) bonds of Ib-d are also regioselectively cleaved by α-CyD (Table 2).

In contrast to the regioselective P-O(2′) cleavages by α-CyD, β- and γ-CyDs promote the cleavages of the P-O(3′) bonds of I, with suppression of

TABLE 2
Selectivities and Rate Constants for the Cleavage of Ib-Id in the Presence and Absence of CyDs[a]

Substrate	CyD	Concentration $(10^{-2} M)$	Rate constant $(10^{-4} min^{-1})$	Selectivity[b] (%)
Ib	α-CyD	1.0	1.7	70
		5.0	1.9	94
	β-CyD	1.0	1.3	43
	γ-CyD	1.0	1.3	50
	None	—	1.3	50
Ic	α-CyD	1.0	4.2	67
		5.0	4.7	76
	None	—	3.5	46
Id	α-CyD	1.0	4.8	62
		5.0	6.2	67
	None	—	2.6	52

[a]   pH 11.08 (bicarbonate buffer), 20°C.
[b]   III/(II + III).

TABLE 3
Selectivities and Rate Constants for the Cleavage of Ic in the Presence and Absence of CyDs[a]

CyD	Concentration $(10^{-2} M)$	Rate constnt $(10^{-4} min^{-1})$	Selectivity[b] (%)
β-CyD	1.0	11.3	85
	1.5	13.3	88
γ-CyD	1.0	5.6	60
α-CyD	1.0	5.2	33
Hepta-2,6-dimethyl-β-CyD	1.0	3.0	47
Hexa-2,6-dimethyl-α-CyD	1.0	2.1	47
None	—	2.8	46

[a]   pH 11.08, 20°C.
[b]   IIc/(IIc + IIIc).

the P-O(2') cleavages (Tables 3 and 4). Thus, the molecular size of the CyD exhibits an overwhelming effect on the regioselective catalyses.

The formation of IIc by the P-O(3') cleavage of Ic is accelerated 6.3-fold by the complex formation with β-CyD, whereas the formation of IIIc is decelerated by about 90%. Thus, the ratio of the rate constant for the formation of IIc to that for the formation of IIIc in the β-CyD-Ic complex is 79, corresponding to the maximal selectivity of 99% for the P-O(3') cleavage.

TABLE 4
Selectivities and Rate Constants for the Cleavage of Ia, Ib, and Id in the Presence and Absence of CyDs[a]

Substrate	CyD	Concentration $(10^{-2} M)$	Rate constant $(10^{-4} \text{ min}^{-1})$	Selectivity[b] (%)
Ia	β-CyD	1.0	1.9	55
	None	—	1.7	53
Ib	β-CyD	1.0	1.3	57
	None	—	1.3	50
Id	β-CyD	1.0	2.3	57
		1.5	3.1	61
	γ-CyD	1.0	3.8	61
	α-CyD	1.0	4.8	38
	None	—	2.6	48

[a] pH 11.08 (bicarbonate buffer), 20°C.
[b] III/(II + III).

TABLE 5
Selectivities for the CyD-Catalyzed Cleavages of RNAs at pH 11.08, 50°C

	Selectivity[a] (%)	
RNA	3'-Terminal fragment by α-CyD (0.15 M)	2'-Terminal fragment by β-CyD (0.05 M)
Poly[A]	77 (51)	86 (49)
Poly[C]	97 (48)	52 (52)
Poly[G]	68 (50)	61 (50)
Poly[U]	97 (49)	49 (51)

[a] The numbers in parentheses show the selectivities in the absence of CyDs.

B. CyD-INDUCED REGIOSELECTIVE CLEAVAGES OF RIBONUCLEOTIDE POLYMERS[14,15]

The regioselective catalyses by CyDs are applicable to the cleavages of the polymers of ribonucleotides (Table 5). In the cleavage of poly[C] in the presence of 0.15 M α-CyD, the formation of the 3'-terminal poly[C] fragments (III-p) is predominant, with selectively of 97% (the rest is the 2'-isomer: II-p). Selective cleavages of poly[U], poly[A], and ply[G] to the corresponding 3'-terminal fragments (III-p) are also successful by α-CyD.

In contrast, the 2'-terminal fragments (II-p) are selectively formed from poly[A] and poly[G] by use of β-CyD as catalyst.

These regioselectivities are virtually identical with those exhibited by CyDs in the cleavages of the 2',3'-cyclic phosphates of the ribonucleosides, which are present in the 3'-side of the cleaved phosphodiester bonds (Tables 1 to 4).

Thus, the cleavages of the polymers are two-step reactions involving the 2′,3′-cyclic monophosphates of the terminal ribonucleotides (I-p in Figure 1), as are the ribonuclease-catalyzed cleavages of RNAs. α-CyD induces the regioselective cleavages of the P-O(2′) bonds of I-p (route A), and β-CyD induces the P-O(3′) cleavages (route B).

C. STRUCTURES OF THE COMPLEXES BETWEEN CyD AND I

On the complex formation between α-CyD and Ia, the resonances for the H-2 and H-3 protons of α-CyD shifted considerably toward the higher magnetic field (0.075 and 0.055 ppm when the charged concentrations of α-CyD and Ia were 10^{-2} and 2×10^{-2} M, respectively). The H-1 and H-4 protons also showed upfield shifts (0.036 and 0.028 ppm). On the other hand, the H-5 and H-6 protons of the cytosine residue of Ia experienced downfield shifts on the complex formation (0.014 and 0.018 ppm when $[\alpha\text{-CyD}]_0 = 2 \times 10^{-2} M$ and $[Ia]_0 = 10^{-2} M$). The H-3′ proton of the ribose residue of Ia also showed a downfield shift (0.018 ppm). The changes for all the other protons were marginal. Formation of the complex between α-CyD and Ib exhibited similar changes in the chemical shifts.

These results indicate that the secondary hydroxyl groups of α-CyD form hydrogen bonds as proton donors, and thus the electron densities on the H-2 and H-3 atoms are increased. Both these protons and the secondary hydroxyl groups are directly attached to the C-2 and C-3 carbon atoms. The upfield shifts for the H-1 and H-4 protons, which are on the adjacent carbon atoms, are ascribed to through-chain transfer of the effects from the C-2 and C-3 carbon atoms.

Thus, the structure of the α-CyD-Ia complex is proposed as depicted in Figure 2a. The phosphate residue of Ia (at the O(1) and O(2) atoms) forms hydrogen bonds with the secondary hydroxyl group of α-CyD, and the oxygen atom at the C-2 carbon of the cytosine residue forms another hydrogen bond with the secondary hydroxyl group of the farthest glucose residue. The plane of the cyclic phosphate residue of Ia is almost parallel to the longitudinal axis of the cavity of α-CyD. As a result, the cavity of α-CyD is flexibly capped by Ia at the secondary hydroxyl side. The structure is stabilized by the cooperation of several types of hydrogen bonds. According to the CPK molecular model study, the proposed complex is formed without steric repulsion. The α-CyD-Ib complex has a structure similar to the α-CyD-IA complex.[16]

Effective formation of the hydrogen-bonding complexes between α-CyD and I in water, a hydrogen-bonding solvent, is ascribed to the considerably small pK_a value (around 12)[7] of the secondary hydroxyl groups of α-CyD. The proton-donating ability of these hydroxyl groups is so large that α-CyD forms complexes with I without significant competition with water. Otherwise, use of a nonhydrogen bonding solvent such as chloroform is absolutely required, as is the case for most of the hydrogen-bonding host compounds previously studied.[17]

A possibility that the cytosine or uracil residue of Ia or Ib is deeply included in the cavity is ruled out by the ¹H-NMR spectroscopy. The chemical shifts of the H-1, H-2, and H-4 protons of α-CyD are hardly affected by the anisotropic shielding effects of the guest compounds included in the cavity, since they are on the exterior of the cavity.

In contrast, the complex formation between β-CyD and Ic or d involves inclusion of the nucleic base in the cavity. Here, the NMR signals for the H-3 and H-5 protons of β-CyD, which are in the interior of the cavity, shifted considerably toward the higher magnetic field (0.045 and 0.070 ppm when the charged concentrations of β-CyD and Ic were 10^{-2} and 2×10^{-2} M, respectively). These upfield shifts are attributable to the anisotropic shielding effects of the adenine residue of Ic which is included in the cavity of β-CyD. The larger shift for the H-5 protons than for the H-3 protons (ratio of 1.6:1.0) shows that the penetration of the adenine residue into the cavity is deep, so that the H-5 protons are efficiently shielded by the ring current effect. Marginal shifts for the other protons, which are in the exterior of the cavity, are consistent with this argument. Thus, the structure of the complex is proposed as schematically depicted in Figure 2b.

D. MECHANISM OF THE REGIOSELECTIVE CATALYSIS BY CyD

The regioselective catalyses by CyDs are definitely attributable to the discrimination between the P-O(2') bonds of I's and the P-O(3') bonds by the formation of the CyD-I complexes. Otherwise, the chemical circumstances of these two bonds are virtually identical.

In the α-CyD-I complexes, the O(3') atoms of I's are located in or near the apolar cavity of the α-CyD (Figure 2a). Thus, the cleavages of the P-O(3') bonds are highly suppressed, since the formation of the O(3') alkoxide ions in the apolar atmosphere are energetically unfavorable. In contrast, the formation of the O(2') alkoxide ions for the P-O(2') cleavages is facilitated by effective solvation of the ions with water molecules.

Furthermore, the efficient P-O(2') cleavages are associated with the fact that the phosphate residues of I's gradually come closer to the secondary hydroxyl groups of α-CyD as the reactions proceed. This strengthens the hydrogen bonds between the phosphate residues and the hydroxyl groups, and stabilizes the transition state. In the P-O(3') cleavage, however, the distance between the two groups gradually increases, and the transition state is destabilized.

In the β-CyD-I complexes, however, the O(2') atoms of I's are located near the cavity. As a result, the P-O(3') bonds are cleaved efficiently due to the same reasons as described above for the α-CyD complexes, except for the replacement of the O(3') atoms with the O(2') atoms and of the O(2') atoms with the O(3') atoms.

Covalent catalysis by CyD, which involves the formation of a covalent intermediate by the nucleophilic attack of the secondary alkoxide ion of CyD

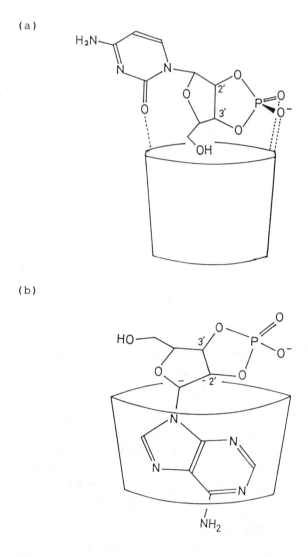

FIGURE 2.   Proposed structures of the α-CyD-Ia complex (a)
and the β-CyD-Ic complex (b).

at the phosphorus atom of I, is unlikely, since the rate of disappearance of I
was identical to the total rate of the appearance of II and III.

No catalytic effects of hexa-2,6-dimethyl-α-CyD, hepta-2,6-dimethyl-β-
CyD, and hepta-2,3,6-trimethyl-β-CyD definitely support the proposed mech-
anisms, which involve essential roles of the secondary hydroxyl groups of
CyDs. Significant dependence of the regiospecific catalyses on the kind of
CyD also agrees with the mechanisms, since they require strict regulation of
the stereochemistry of the CyD complexes for the effective catalyses.

E. EFFECTS OF COMPETITIVE INHIBITOR ON THE CyD-INDUCED REGIOSELECTIVE CLEAVAGE OF I

These arguments are confirmed by the results of the competitive inhibition experiments. 4-Nitrophenolate as competitive inhibitor decreases the rate constant and the selectivity in both the α-CyD- and β-CyD-induced cleavages. The magnitudes of the decrease for the β-CyD-induced cleavage of Ic satisfactorily agreed with the values evaluated under the assumption that the complex between β-CyD and 4-nitrophenolate has no catalytic activity. The cavity of β-CyD is absolutely necessary for the catalysis, as indicated by the above mechanism.

In the α-CyD-induced cleavage of Ia, however, the magnitudes of the decrease in both the rate constant and the selectivity are considerably smaller than the values evaluated under the same assumption. Thus, the α-CyD-4-nitrophenolate complex still has some catalytic activity. This fact confirms the above argument that the α-CyD-Ia complex is not an "inclusion" complex, but, rather, a hydrogen-bonding one, and that the catalysis by α-CyD does not require the cavity. If the regioselective catalysis involved deep penetration of Ia in the cavity, the catalytic activity would be totally inhibited by the inhibitor.

IV. CONCLUSION

The regioselective cleavages of the 2',3'-cyclic monophosphates of ribonucleosides are achieved by CyDs as catalysts. α-CyD promotes the cleavages of the P-O(2') bonds, whereas β- and γ-CyDs enhance the P-O(3') cleavages. These catalyses are ascribed to the formation of the complexes between the cyclic phosphates and CyDs. The circumstances of the O(2') atoms and the O(3') atoms are largely differentiated by the complex formation. The polymers of ribonucleotides are also regioselectively cleaved by CyDs.

The regiospecificity of β- and γ-CyDs is the reverse of the specificity (the P-O(2') cleavage) of ribonuclease, whereas the specificity of α-CyD is identical to that of the enzyme. The remarkable dependence of regiospecificity on the kind of CyD comes from the difference of the structure of the CyD-substrate complex (the inclusion type for β- and γ-CyDs, and the hydrogen bonding type for α-CyD).

The present finding indicates that cyclodextrin chemistry, which was enormously and beautifully developed by Professor Myron L. Bender, can be even more fruitful in the future.

REFERENCES

1. **Adams, R. L. P., Knowler, J. T., and Leader, D. P.,** *The Biochemistry of the Nucleic Acids,* 10th ed., Chapman and Hall, London, 1986.
2. **Dugas, H. and Penney, C.,** *Bioorganic Chemistry,* Springer-Verlag, New York, 1981.
3. **Richards, F. M. and Wyckoff, H. W.,** *Enzymes,* Vol. 4, 3rd ed., Boyer, P. D., Ed., Academic Press, New York, 1971, chap. 24.
4. **Breslow, R., Doherty, J. B., Guillot, G., and Lipsey, C.,** *J. Am. Chem. Soc.,* 100, 3227, 1978.
5. **Breslow, R., Bovy, P., and Hersh, C. L.,** *J. Am. Chem. Soc.,* 102, 2115, 1980.
6. Breslow reported that chemically modified CyDs, in which two imidazolyl residues are introduced, exhibit regiospecific cleavage of cyclic monophosphate of 4-*tert*-butylca-techol; see References 4 and 5.
7. For a review on CyD: **Bender, M. L. and Komiyama, M.,** *Cyclodextrin Chemistry,* Springer-Verlag, Berlin, 1978.
8. **Komiyama, M.,** *Chem. Lett.,* 689, 1988.
9. **Komiyama, M.,** *Chem. Lett.,* 1121, 1988.
10. **Komiyama, M.,** *Makromol. Chem. Rapid Commun.,* 9, 453, 1988.
11. **Komiyama, M.,** *J. Am. Chem. Soc.,* 111, 3046, 1989.
12. **Komiyama, M.,** *Makromol. Chem. Rapid Commun.,* 9, 453, 1988.
13. **Komiyama, M. and Takeshige, Y.,** *J. Org. Chem.,* 54, 4936, 1989.
14. **Komiyama, M.,** *Polym. J.,* 20, 525, 1988.
15. **Komiyama, M.,** *Carbohydr. Res.,* 192, 97, 1989.
16. In aqueous solutions, Ia and Ib overwhelmingly take syn conformations with respect to the rotation of the *N*-glycosyl bonds: **Lavallee, D. K. and Coulter, C. L.,** *J. Am. Chem. Soc.,* 95, 576, 1973; **Lapper, R. D. and Smith, I. C. P.,** *J. Am. Chem. Soc.,* 95, 2880, 1973.
17. **Rebek, J. Jr., Askew, B., Ballester, P., Buhr, C., Jones, S., Nemeth, D., and Williams, K.,** *J. Am. Chem. Soc.,* 109, 5033, 1987; **Hamilton, A. D. and Van Engen, D.,** *J. Am. Chem. Soc.,* 109, 5035, 1987.

Chapter 8

ESTERASE-LIKE ACTIVITY AND DRUG BINDING OF HUMAN SERUM ALBUMIN

Yukihisa Kurono and Ken Ikeda

TABLE OF CONTENTS

I. INTRODUCTION

Human serum albumin (HSA) is the most abundant protein in blood plasma and the transport protein for many endogenous and exogenous compounds.[1-3] Because the binding of drugs to HSA has important consequences for the intensity of biological activity and for the pharmacokinetic behavior of drugs, there have been a large number of studies on drug-HSA interactions.[4,5] Although many quantitative studies (e.g., determination of the binding constants and the number of binding sites) have been done, there have been only a few studies on the location of the binding sites.[6,7] Since 1975 when the complete amino acid sequence of HSA was reported,[8,9] the studies on the localization of the drug binding sites and on the classification of drugs with respect to the specific binding sites have been carried out from the viewpoint of drug displacement *in vivo*. The methodologies used are fluorescence spectroscopy,[10] chemical modification of the binding sites,[11] usage of immobilized HSA,[12] etc. We have developed a novel kinetic method using the competitive inhibition of the esterase-like activity of HSA. A similar kinetic method was also concurrently studied by Means et al.[13]

In this chapter, we describe some esterase-like behavior of HSA, the distinction and identification of the drug-binding sites based on the inhibition patterns of the activity, and the determination of the drug-binding parameters by analog computer.

II. ESTERASE-LIKE ACTIVITY OF HSA

HSA (Sigma Chemical Co., Fraction V) was used after purification by Chen's method.[14] Substrates (S) were purchased commercially and/or synthesized as necessary. The reaction of S with HSA (excess concentration compared with S) in the absence and presence of drug was followed, using a spectrophotometer and stopped-flow spectrophotometer. The pseudo first-order rate constant (k_{obs}) was determined from a plot of $\log|A_\infty - A_t|$ against time, where A_∞ and A_t are the absorbances at the completion of the reaction and at time t, respectively.

The k_{obs} values for the reaction of *p*-nitrophenyl acetate (S, NPA) with HSA increase hyperbolically with the HSA concentration increase, suggesting saturation kinetics for the reaction as shown in Scheme 1.

$$S + HSA \overset{K_S}{\leftrightarrows} S \cdot HSA \overset{k_2}{\longrightarrow} p\text{-nitrophenol} + acyl\text{-}HSA$$
$$\downarrow k_0$$

p-nitrophenol + acetic acid

SCHEME 1

In Scheme 1, S · HSA is the Michaelis-Menten-type complex between S and HSA, and K_S is the dissociation constant of S · HSA. The catalytic rate constant and the spontaneous hydrolysis rate constant of S are expressed by k_2 and k_0, respectively. According to Scheme 1, the k_{obs} value obtained experimentally can be represented by Equation 1.[15,16]

$$k_{obs} = \frac{k_0 K_S + k_2 [HSA]_0}{K_S + [HSA]_0} \qquad (1)$$

where $[HSA]_0$ is the initial concentration of HSA. The K_S and k_2 values can be calculated from the slope and intercept of the double-reciprocal plot based on Equation 2.[15-17]

$$\frac{1}{k_{obs} - k_0} = \frac{K_S}{(k_2 - k_0)} \cdot \frac{1}{[HSA]_0} + \frac{1}{k_2 - k_0} \qquad (2)$$

At pH 7.0 and 25°C, for example, the K_S, k_2, and k_0 values were 1.55×10^{-4} M, 2.93×10^{-2} s^{-1}, and 2.29×10^{-5} s^{-1}, respectively.[15,16] The k_2/k_0 value is 1.28×10^3, indicating a large esterase activity of HSA. To elucidate the reactivities of HSA toward esters, the quantitative structure-activity relationships for the reactions with phenyl acetates (number of compounds (n = 5) and p-nitrophenyl esters (n = 8) were examined kinetically.[15]

Chemical modification of the tyrosine-411 (Tyr-411) residue of HSA with diisopropylfluorophosphate (DFP) inhibited the reaction with NPA, indicating the catalytic group to be Tyr-411.[18-20] The reaction of NPA with HSA is competitively inhibited by many drugs and compounds (see the next section),[13,19,21,22] showing the enzyme-like activity of HSA. We named this reactive, drug-binding site the R-site.

To elucidate further the details of the esterase activity of HSA, the reaction of N-trans-cinnamoylimidazoles with HSA was investigated kinetically at various pHs at 25°C.[23] The reaction consists of the acylation (k_2) of HSA (probably at the Tyr-411 residue) by the substrate (CI) and the deacylation (k_3) of cinnamoyl-HSA, as shown in Scheme 2.

$$CI + HSA \overset{K_S}{\rightleftharpoons} CI \cdot HSA \overset{k_2}{\longrightarrow} \text{imidazole} + \text{cinnamoyl-HSA}$$
$$\downarrow k_0 \qquad\qquad\qquad\qquad\qquad\qquad\qquad \downarrow k_3$$
$$\text{cinnamic acid} + \text{imidazole} \qquad\qquad \text{cinnamic acid} + HSA$$

SCHEME 2

The pH profiles of the kinetics parameters for the reaction with cinnamoyl-imidazole (CI) are shown in Figure 1. The k_2 value is ~10- to 100-fold larger than k_0 over the pH range examined. The deacylation (k_3) is, however, slower

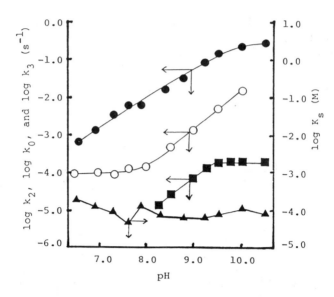

FIGURE 1.   The pH profiles of kinetic parameters for the reaction of CI with HSA at 25°C. ●, log k_2; ■, log k_3; ○, log k_0; ▲, log K_S.

than the hydrolysis (k_0) of CI. In this context, esterase-like activity rather than the intrinsic esterase may be an appropriate expression for the activity of HSA toward the amide and ester substrates.

HSA also was found to have the esterase-like activity toward nitroaspirins. Acetylation of the lysine-199 (Lys-199) residue of HSA with aspirin[20,24,25] retarded the reactions of 5-nitroaspirin (NA) and 3,5-dinitroaspirin (DA) with HSA. The complex formation between NA and the reactive site of HSA decreased the fluorescence intensity (quenching) due to tryptophan-214 (Trp-214) residue.[19,26,27] These results indicate that the reactive site (named the U-site) toward aspirin derivatives locates near the Lys-199 and Trp-214 residues of HSA. To characterize the U-site and to elucidate the reactivity, as in the case of the R-site, the reactions of substituted aspirins (n = 5) and 5-nitrosalicyl esters (n = 6) with HSA were investigated kinetically.[28]

Enantioselectivity for the reaction of D(L)-*p*-nitrophenyl α-methoxyphenyl acetate with HSA was observed.[29] HSA reacts with the D-enantiomer about threefold faster than with the L-enantiomer, mainly due to the catalytic step, and not the binding step. HSA also has esterase-like activity toward organic phosphate, 2,4-dinitrophenyl diethyl phosphate.[30] The catalytic group toward this substrate was suggested to be an imidazole group of histidine residue near the Tyr-411 residue (R-site).

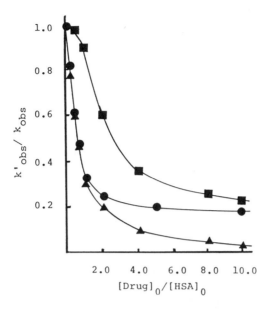

FIGURE 2.   Effects of drugs on the reaction rate of NPA
with HSA. ●, clofibric acid (R-type inhibition); ▲, flufenamic acid (R-T-type inhibition); ■, phenylbutazone
(U-R-type inhibition). $[NPA]_0 = 1.00 \times 10^{-5} M$; $[HSA]_0$
$= 5.00 \times 10^{-5} M$; pH 7.4 phosphate buffer containing
0.5% (v/v) acetonitrile and 25°C; k_{obs} for NPA $= 6.20 \times 10^{-3} s^{-1}$.

III. INHIBITION PATTERNS FOR ESTERASE-LIKE ACTIVITIES OF HSA

Figure 2 shows three typical inhibition patterns for the reaction of NPA
with HSA. In this figure, k'_{obs} on the ordinate is the rate constant in the
presence of drug. These results can be explained by the following three
mechanisms.[19]

R-type inhibition — Drugs shown by closed circles bind exclusively to
the primarily reactive site (R-site) toward NPA, which corresponds to Sudlow's Site 2.[10] Site 2 drugs (clofibric acid, ibupofen, etc.) show this inhibition
pattern.

R-T-type inhibition — Drugs shown by closed triangles bind not only to
the R-site, but also to the secondarily reactive site toward NPA (named the
T-site), whose location has not been found yet. Flufenamic acid and ethra-
crynic acid, which also belong to the Site 2 drugs, show this inhibition pattern.

U-R-type inhibition — Drugs shown by closed squares do not bind primarily to the R-site. In the initial shoulder region of the curve, the drug binds
mainly to its primary binding site, which was found to be the U-site corresponding to Sudlow's Site 1. With further increase in the drug concentration

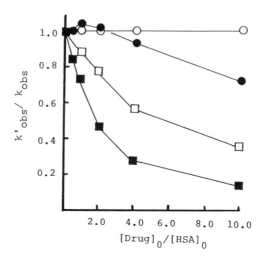

FIGURE 3. Effects of phenylbutazone and clofibric acid on the reaction rates with nitroaspirins (NA and DA). ■, phenylbutazone for NA: □, phenylbutazone for DA; ●, clofibric acid for NA; ○, clofibric acid for DA. The experimental conditions are the same as those in Figure 2; k_{obs} for NA = 4.94×10^{-4} s^{-1}; k_{obs} for DA = 2.30×10^{-1} s^{-1}.

(after the shoulder region), the drug binds gradually to the R-site. Most of Sudlow's Site 1 drugs show this U-R type of inhibition.

Figure 3 shows the inhibition patterns of phenylbutazone and clofibric acid for the reactions of nitroaspirins (NA and DA) with HSA.[27] Quite contrastive inhibition patterns are obtained between nitroaspirins (Figure 3) and NPA (Figure 2). Phenylbutazone directly inhibits (without shoulders) the reactions with nitroaspirins, and clofibric acid does not inhibit the reactions with nitroaspirins at a low $[drug]_0/[HSA]_0$ ratio. The difference between the inhibition curves of NA (closed symbols) and DA (open symbols) for each drug (phenylbutazone or clofibric acid) seems to be due to the difference in the dissociation constant (K_S) of the substrate · HSA complexes, i.e., the K_S for the reaction of NA with HSA is larger than that for DA. These inhibition patterns support the above explanation for the distinction between the R- and U-sites, and for the U-site as the reactive site toward aspirin derivatives.

NPA and nitroaspirins, therefore, can be used as probes for identifying and distinguishing between the specific drug binding sites on HSA, i.e., the R- and U-sites. We have examined many drugs such as sulfonamides, sulfonylureas, benzodiazepines, penicillins, cephalosporins, phenothiazines, cardiac glycosides, etc.[31,32] Table 1 shows the binding site nomenclatures used by several investigators. The results obtained in our studies are in agreement with those obtained by various methods, and support their results from a quite different standpoint. Means et al.[13] also carried out similar kinetic studies using only NPA and obtained results similar to ours.

TABLE 1
Nomenclatures for Drug Binding Sites of Human Serum Albumin

	Fluorescence spectroscopy (Sudlow et al., Australia[10])	Immobilized HSA (Sjoholm et al., Sweden[12])	Chemical modification (Muller et al., Germany[11])	Kinetics (Kurono et al., Japan[19,27])	Location on amino acid sequence of HSA[13,19,27]
Clofibric acid, benzodiazepines, indole analogs, etc.	Site 2 (dansyl-sarcosine)	Site 1 (diaze-pam)	Diazepam site	R-site (NPA)	Tyr-411
Phenylbutazone, warfarin, pheny-toin, etc.	Site 1 (dansyl-amide)	Site 3 (war-farin)	Warfarin site	U-site (ni-troaspirins)	Lys-199 and Trp-214

Note: The drugs or compounds in parentheses are probes used for the competitive displacement.

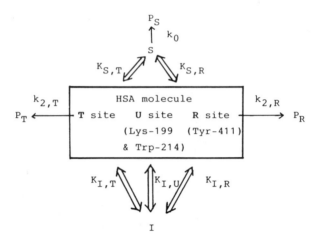

FIGURE 4.   Relationship between the esterase-like active sites toward NPA (S) and the drug binding sites on HSA. See text for the explanation of each parameter.

IV. DETERMINATION OF KINETIC AND BINDING PARAMETERS BY ANALOG COMPUTER

The relationship between the esterase-like active site toward NPA (S) and the drug binding sites on HSA is shown schematically in Figure 4,[33] where $K_{S,R}$ and $K_{S,T}$ are the dissociation constants between S and the esterase-like active sites (R- and T-sites), respectively. $K_{I,R}$, $K_{I,U}$, and $K_{I,T}$ are the dissociation constants between drug (inhibitor, I) and the corresponding binding sites (R-, U-, and T-sites). $k_{2,R}$, $k_{2,T}$, and k_0 show the first-order rate constants of the complexes between S and the reactive sites, and of the uncatalyzed reaction of S, respectively. P_R, P_T, and P_S represent the products from the reactions characterized by $k_{2,R}$, $k_{2,T}$, and k_0, respectively, all of which are *p*-nitrophenol. Out of many parameters shown in Figure 4 ($K_{S,R}$, $K_{S,T}$, $k_{2,R}$, $k_{2,T}$, and k_0), the kinetic parameters for the reactions of S with HSA were determined using clofibric acid as an inhibitor. Clofibric acid distinguished between the two sites, the R-site being inhibited by clofibric acid whereas the T-site was not. The parameters obtained are listed in Table 2.[22] It was difficult, however, to determine the dissociation constants ($K_{I,R}$, $K_{I,T}$, and $K_{I,U}$) between drug (I) and the individual binding sites by the conventional analytical method because of the complicated inhibition system. So we used an analog computer to determine those parameters.[33] The details of the computer programming are described in the literature.[33,34] Figure 5 shows an example of the results of computer simulation on the U-R type inhibition curve. From this curve, the $K_{I,U}$ and $K_{I,R}$ values for phenylbutazone can be estimated as 1×10^{-6} and 3×10^{-5} *M*, respectively. We also simulated both the R and R-T types of inhibition (such as the solid curves in Figure 2),

TABLE 2
Kinetic Parameters in Figure 4 for the R-Type Inhibition and Specificity Constants for the R- and T-Sites

Kinetic parameter	Value	Kinetic parameter	Value
$k_{2,R}(s^{-1})$	3.14×10^{-2}	$K_{I,R}$ (M)	4.65×10^{-6}
$k_{2,T}(s^{-1})$	9.07×10^{-3}	$k_R = k_{2,R}/K_{S,R}$ $(M^{-1}\,s^{-1})$	2.36×10^2
k_0 (s^{-1})	1.67×10^{-5}	$k_T = k_{2,T}/K_{S,T}$ $(M^{-1}\,s^{-1})$	2.94×10
$K_{S,R}$ (M)	1.33×10^{-4}	$100\,k_T/(k_R + k_T)$ $(\%)$	11.1
$K_{S,T}$ (M)	3.09×10^{-4}		

Note: 0.067 M phosphate buffer (μ = 0.2) containing 0.5% (v/v) acetonitrile at pH 7.4 and 25°C.

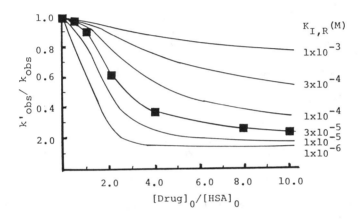

FIGURE 5.   U-R-type inhibition curves at different $K_{I,R}$ values and constant $K_{I,U}$ value (1 × 10^{-6} M). $[HSA]_0 = [HSA_R]_0 = [HSA_T]_0 = [HSA_U]_0 = 5 \times 10^{-5}\,M$; $[S]_0 = 1 \times 10^{-5}\,M$; $k_{obs} = 6.20 \times 10^{-3}\,s^{-1}$. ■, experimental results for phenylbutazone; conditions are the same as those in Figure 2.

and determined the dissociation constants for all the drugs mentioned in section III. From the curves in Figure 2, the following values were obtained: $K_{I,R} = 4 \times 10^{-6}\,M$ for clofibric acid; $K_{I,R} = 2 \times 10^{-6}\,M$ and $K_{I,T} = 1 \times 10^{-4}\,M$ for flufenamic acid. The binding parameters thus determined correspond well to those obtained by conventional methods such as equilibrium dialysis, etc.

REFERENCES

1. **Rosenoer, V. M., Oratz, M., and Rothschild, M. A.,** Eds., *Albumin Structures, Function and Uses,* Pergamon Press, Oxford, 1977.
2. **Peters, T., Jr.,** *Adv. Protein Chem.,* 37, 161, 1985.
3. **Rothschild, M. A., Oratz, M., and Schreiber, S. S.,** *Hepatology,* 8, 385, 1988.
4. **Vallner, J. J.,** *J. Pharm. Sci.,* 66, 447, 1977.
5. **Kragh-Hansen, U.,** *Pharmacol. Rev.,* 33, 17, 1981.
6. **Fehske, K. J., Muller, W. E., and Wollert, U.,** *Biochem. Pharmacol.,* 30, 687, 1981.
7. **Brown, J. R. and Shockley, P.,** Serum albumin: structure and characterization of its ligand binding sites, in *Lipid-Protein Interactions,* Vol. 1, Jost, P. C. and Griffith, O. H., Eds., John Wiley & Sons, New York, 1982, 25.
8. **Behrens, P. Q., Spiekerman, A. M., and Brown, J. R.,** *Fed. Proc.,* 34, 591, 1975.
9. **Melown, B., Moravek, L., and Kostka, V.,** *FEBS Lett.,* 58, 134, 1975.
10. **Sudlow, G., Birkett, D. J., and Wada, D. N.,** *Mol. Pharmacol.,* 11, 824, 1975; Sudlow, G., Birkett, D. J., and Wada, D. N., *Mol. Pharmcol.,* 12, 1052, 1976.
11. **Fehske, K. J., Muller, W. E., and Wollert, U.,** *Biochem. Biophys. Acta,* 577, 346, 1979; **Fehske, K. J., Muller, W. E., and Wollert, U.,** *Arch. Biochem. Biophys.,* 205, 217, 1980; **Fehske, K. J., Schlafer, V., Wollert, U., and Muller, W. E.,** *Mol. Pharmacol.,* 21, 387, 1981.
12. **Sjoholm, I., Ekman, B., Kober, A., Ljungstedt-Pahlman, Seiving, B., and Sjodin, T.,** *Mol. Pharmacol.,* 16, 767, 1979; **Kober, A. and Sjoholm, I.,** *Mol. Pharmacol.,* 18, 421, 1980.
13. **Means, G. E., Sollenne, N. P., and Mohammed, A.,** *Adv. Chem. Ser.,* 198, 325, 1982.
14. **Chen, R. F.,** *J. Biol. Chem.,* 242, 173, 1967.
15. **Kurono, Y., Maki, T., Yotsuyanagi, T., and Ikeda, K.,** *Chem. Pharm. Bull.,* 27, 2781, 1979.
16. **Ikeda, K., Kurono, Y., Ozeki, Y., and Yotsuyanagi, T.,** *Chem. Pharm. Bull.,* 27, 80, 1979.
17. **Lineweaver, H. and Burk, D.,** *J. Am. Chem. Soc.,* 56, 658, 1934.
18. **Sanger, F.,** *Proc. Chem. Soc.,* 76, 1963.
19. **Ozeki, Y., Kurono, Y., Yotsuyanagi, T., and Ikeda, K.,** *Chem. Pharm. Bull.,* 28, 535, 1980.
20. **Brown, J. R.,** Serum albumin: amino acid sequence, in *Albumin Structure, Function and Uses,* Rosenoer, V. M., Oratz, M., and Rothschild, M. A., Eds., Pergamon Press, Oxford, 1977, 27.
21. **Koh, S.-W. M., and Means, G. E.,** *Arch. Biochem. Biophys.,* 192, 73, 1979.
22. **Kurono, Y., Ohta, N., Yotsuyanagi, T., and Ikeda, K.,** *Chem. Pharm. Bull.,* 29, 2345, 1981.
23. **Ohta, N., Kurono, Y., and Ikeda, K.,** *J. Pharm. Sci.,* 72, 385, 1983.
24. **Walker, J. E.,** *FEBS Lett.,* 66, 173, 1976.
25. **Gambhir, K. K., McMenamy, R. H., and Watson, F. J.,** *Biol. Chem.,* 250, 6711, 1975.
26. **Chignell, C. F.,** *Mol. Pharmacol.,* 6, 1, 1970.
27. **Kurono, Y., Yamada, H., and Ikeda, K.,** *Chem. Pharm. Bull.,* 30, 296, 1982.
28. **Kurono, Y., Yamada, H., Hata, H., Okada, Y., Takeuchi, T., and Ikeda, K.,** *Chem. Pharm. Bull.,* 32, 3715, 1984.
29. **Kurono, Y., Kondo, T., and Ikeda, K.,** *Arch. Biochem. Biophys.,* 227, 339, 1983.
30. **Yoshida, K., Kurono, Y., Mori, Y., and Ikeda, K.,** *Chem. Pharm. Bull.,* 33, 4995, 1985.
31. **Ikeda, K., Kurono, Y., Ozeki, Y., Ohta, N., and Yotsuyanagi, T.,** *J. Pharmacobio-Dyn.,* 3(9), s1, 1980.

32. **Kurono, Y. and Ikeda, K.,** *J. Pharmacobio-Dyn.,* 5(4), s50, 1982.
33. **Kurono, Y. and Ikeda, K.,** *Chem. Pharm. Bull.,* 29, 2993, 1981.
34. **Roberts, D. V.,** *Enzyme Kinetics,* Cambridge University Press, London, 1977, 254.

Chapter 9

IDENTIFICATION AND CHARACTERIZATION OF REACTION INTERMEDIATES IN THE CHEMICAL AND ENZYMATIC ISOMERIZATION OF β,γ-UNSATURATED KETONES TO α,β-UNSATURATED KETONES

Ralph M. Pollack

TABLE OF CONTENTS

I. INTRODUCTION

It seems particularly appropriate to discuss the identification of reaction intermediates in a volume dedicated to the memory of Myron L. Bender. Some of Myron's earliest, and perhaps most significant, work was concerned with the elucidation of the nature of obligatory intermediates in both organic and enzymatic reactions. His use of oxygen-18 to probe the mechanism of ester and amide hydrolysis was inspired, and the identification of a tetrahedral intermediate in these reactions is classic.[1] In the area of enzymology, Myron's kinetic studies on the serine proteases were instrumental in demonstrating that enzyme mechanisms are amenable to study by the methods of physical organic chemistry. His investigations of acyl-enzyme intermediates[2] form the basis for the discussion of the mechanism of serine proteases in virtually every modern biochemistry textbook.

In our laboratory, we have been concerned with the isomerization of β,γ-unsaturated ketones to their α,β-unsaturated isomers (Equation 1).[3]

$$\tag{1}$$

This reaction is catalyzed by primary amines, acids, bases, and the enzyme 3-oxo-Δ^5-steroid isomerase (also called ketosteroid isomerase, KSI). Although the equilibrium constant for this reaction generally favors the conjugated α,β-unsaturated product, exceptions exist. We have investigated the isomerization catalyzed by primary amines, hydroxide ion, and steroid isomerase, and in each case we have been able to monitor directly the reaction of an intermediate in the reaction pathway. In this brief review, the mechanism of catalysis of the isomerization will be presented for each type of catalyst, along with a description of the methods used to identify the intermediate in the reaction and the conclusions that can be drawn from these studies.

II. PRIMARY AMINE CATALYSIS (SCHIFF BASE INTERMEDIATES)

Bender[4] and others (notably Hine[5]) have shown that primary amines are excellent catalysts for ketone enolization through the formation of an intermediate Schiff base. It seemed reasonable to us that primary amines should be able to catalyze the isomerization of β,γ-unsaturated ketones through a similar mechanism, involving an enolization/ketonization sequence (Scheme 1). We initially examined the reaction of 3-methyl-3-cyclohexenone (1) to 3-methyl-2-cyclohexenone (5) catalyzed by 2,2,2-trifluoroethylamine (TFEA, R = CF_3CH_2).[6] The appearance of 5 was monitored spectrally at 240 nm. At concentrations of TFEA < 0.4 M, the formation of 5 is pseudo first-order. However, a significant induction period for the production of 5 suggested the formation of an intermediate. Rapid spectral scans showed the transient accumulation of a species with an absorbance maximum of 268 nm, which decays to the final product (**5**) with a rate constant identical to that obtained from the reaction of **1** to **5**.

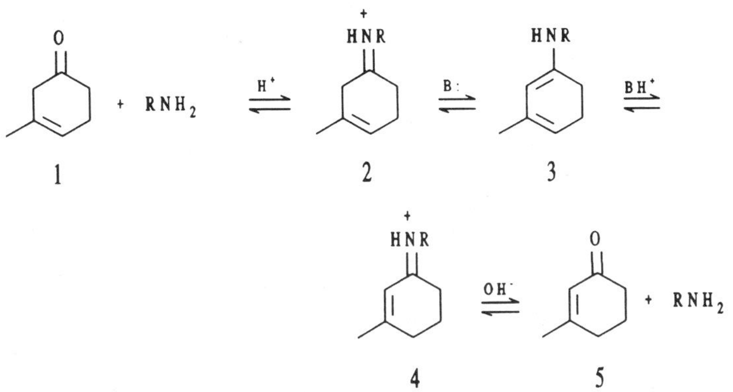

SCHEME 1

The spectral properties of this intermediate suggested that it is the protonated α,β-unsaturated Schiff base **4**. In order to confirm the identity of the intermediate, we synthesized authentic **4** by the reaction of TFEA with **1** in carbon tetrachloride. Upon dissolution of **4** in slightly acidic aqueous solution, a UV spectrum identical to that for the intermediate in the amine-catalyzed isomerization was observed. Hydrolysis of **4** to the α,β-unsaturated ketone **5** occurs with a rate constant indistinguishable from that for the intermediate in the reaction of **1** to **5**. In addition, by monitoring the isomerization in DMSO/D$_2$O (70:30) by nuclear magnetic resonance (NMR), we were able to show that the ^{1}H NMR spectrum of the intermediate is identical to that of **4**.

The overall catalytic efficiency of TFEA in the isomerization of **1** to **5** is limited by the rate of hydrolysis of **4**, since the actual isomerization step **2** → **4** is about 100-fold faster than the hydrolysis of **4**. We examined the hydrolysis of **4** in some detail in order to determine what conditions would lead to an increase in the rate.[7] Two interesting results emerged from these studies: (1) the reaction is general base catalyzed and (2) the hydrolysis rate is enhanced in solutions of low water concentration. The latter result is quite surprising since attack of water on the protonated Schiff base is an integral part of the reaction mechanism.

The hydrolysis of **4** shows kinetics that are similar to those observed for Schiff base derived from aromatic or saturated aliphatic carbonyl compounds, suggesting that a similar mechanism applies. There are kinetic terms for attack by water, for attack by hydroxide ion, and for general base-catalyzed attack of water on the protonated Schiff base. At $0 < $ pH $ < 7$, addition of water is the rate-limiting step (Equation 2); at pH values between 1 and 6, the reaction is buffer catalyzed with a β-value of 0.4 (Equation 3); at pH > 7, the rate-limiting step is attack of hydroxide ion on **4**.

$$
\begin{matrix} H \\ \diagdown \\ H \diagup \end{matrix} O \;+\; \begin{matrix} \diagdown \\ \diagup \end{matrix}C = \overset{+}{N}HR \quad\longrightarrow\quad \begin{matrix} H \\ \diagdown \\ H \diagup \end{matrix}\overset{+}{O} - \overset{|}{\underset{|}{C}} - NHR \tag{2}
$$

$$
B : \curvearrowright H - \underset{\underset{H}{|}}{O} \curvearrowright \begin{matrix} \diagdown \\ \diagup \end{matrix}C = \overset{+}{N}HR \quad\longrightarrow\quad BH^{+} \;+\; HO - \overset{|}{\underset{|}{C}} - NHR \tag{3}
$$

When the solvent is changed from pure water to aqueous solutions of increasing dioxane content, the rate of hydrolysis increases substantially, in spite of the lower concentration of the attacking nucleophile (Table 1). For example, in 0.01 *M* HCl, the rate constant for hydrolysis of **4** is 18-fold greater in 90% dioxane than it is in water. Since the concentration of water has been diminished tenfold, the actual rate constant is 180-fold greater in 90% dioxane than in water. We attributed this result to preferential solvation of oxygen acids relative to nitrogen acids by dioxane.[7]

An even larger solvent effect is observed for the general base-catalyzed addition of water. When the solvent is changed from water to 70% dioxane, the rate constant for the chloroacetate-catalyzed attack of water on **4** is increased by almost 250-fold (Table 2). whereas the same change in solvent only produces an 11-fold increase in the rate constant for the uncatalyzed attack of water. Significantly, the combined effect of general base catalysis by 1 *M* chloroacetate in 70% dioxane compared to the rate in water alone (about 250-fold) is substantially greater than the product of the two individual effects (2.5-fold for 1 *M* chloroacetate in water and 11-fold for the solvent effect on the uncatalyzed rate). The synergism of these two effects no doubt is due to destruction of charges in the reaction with chloroacetate (Equation 4) vs. only charge dispersal for uncatalyzed attack by water (Equation 5). These results suggest that formation and hydrolysis of Schiff base intermediates in enzymatic reactions may be facilitated by a combination of general base catalysis and a hydrophobic active site.

$$
RCOO^{-} \curvearrowright H - \underset{\underset{H}{|}}{O} \curvearrowright \begin{matrix} \diagdown \\ \diagup \end{matrix}C = \overset{+}{N}HR \;\longrightarrow\; \left[RCOO^{\delta-} - - H - - O - - C \overset{\delta+}{=} NHR \atop \underset{H}{|} \right]^{\ddagger} \tag{4}
$$

$$
H - \underset{\underset{H}{|}}{O} \curvearrowright \begin{matrix} \diagdown \\ \diagup \end{matrix}C = \overset{+}{N}HR \;\longrightarrow\; \left[\delta+ \overset{H}{\underset{H}{\diagdown}} O - - C \overset{\delta+}{=} NHR \right]^{\ddagger} \tag{5}
$$

TABLE 1

**Solvent Effects on the Hydrolysis of 2,2,2-Trifluoro-
N-(3-methyl-2-cyclohexenylidene)ethylamine in
Dioxane-Water Solutions at 25° C**

% Dioxane	k_{obs} (0.1 M HCl), s^{-1}	k_{obs} (0.01 M HCl), s^{-1}
0	3.47×10^{-4}	3.61×10^{-4}
20	7.08×10^{-4}	
50	2.13×10^{-3}	2.27×10^{-3}
60	2.83×10^{-3}	
70	3.70×10^{-3}	4.05×10^{-3}
80	4.37×10^{-3}	
90	3.92×10^{-3}	6.43×10^{-3}

Reprinted with permission from Brault, M. and Pollack, R. M., *J. Am. Chem. Soc.*, 98, 247, 1976. Copyright 1976, American Chemical Society.

TABLE 2

**Catalytic Constants for the
Chloroacetate-Catalyzed
Hydrolysis of 2,2,2-Trifluoro-
N-(3-methyl-2-cyclohexenyli-
dene)ethylamine at 25°C**

Solvent	k^B, $M^{-1} s^{-1}$
Water	5.3×10^{-4}
50% dioxane	2.2×10^{-2}
70% dioxane	1.3×10^{-1}

III. ENOL INTERMEDIATES

In the absence of amines, the isomerization of β,γ-unsaturated ketones proceeds through either a dienol (acid catalysis)[8] or a dienolate ion (base catalysis.)[8f,9] Since enols in general[10,11] and dienols in particular[11f,12] are usually quite unstable, with lifetimes in aqueous solution of less than 1 min, it is difficult to examine the reactions of these intermediates in the isomerization reaction directly. Although the vast majority of enols are impossible to isolate, the steroidal trienol 3-hydroxy-3,5,7-estratrien-17-one (**6**) can be synthesized and isolated.[13] This trienol can ketonize by protonation at C-4, C-6, or C-8 to produce the $\Delta^{5,7}$, $\Delta^{4,7}$, or $\Delta^{4,6}$ isomers (**7**, **8**, and **9**, respectively), and thus would be the intermediate in the isomerizations of **7** ⇌ **8**, **7** ⇌ **9**, and **8** ⇌ **9** (Scheme 2).

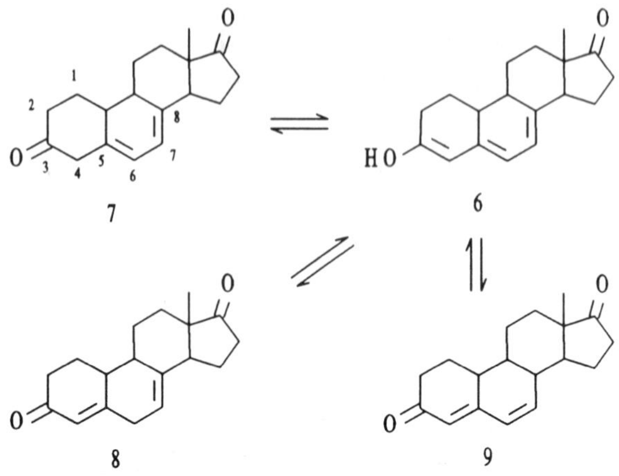

SCHEME 2

We examined the ketonization reaction of **6** catalyzed by acid and by base, as well as the neutral reaction (Figure 1).[14] The reaction mechanism for acid catalysis involves direct protonation by hydronium ion at a carbon-carbon bond to produce a protonated ketone, followed by rapid loss of the proton on the oxygen to yield **7** and **8** in approximately equal amounts. In neutral solution, protonation occurs by hydronium ion on the trienolate ion (**6$^-$**) to give **7** and **8**, whereas in basic solution protonation of the anion by water yields **8** and an unidentified oxidation product. Significantly, none of the most stable isomer (**9**) is formed at any pH value investigated (between 1 and 8).

In other work,[15] we were able to demonstrate the intermediacy of a dienolate ion in the hydroxide ion-catalyzed isomerization of 5-androstene-3,17-dione (**10**) to 4-androstene-3,17-dione (**11**) (Scheme 3). When the reaction is monitored at the absorbance maximum of **11** (248 nm), excellent pseudo first-order kinetics are observed, and a plot of log k_{obs} vs. [OH$^-$] exhibits saturation behavior with increasing hydroxide ion concentration. The observed rate constants fit the rate expression for the mechanism of Scheme 3 with a value of K ($= k_1/k_{-1}$) of 12 ± 2 M^{-1}. This value can be converted to an aqueous pK_a of 12.72 ± 0.08 for **10**, which requires that the intermediate anion **12** be formed in significant quantities during the reaction at relatively high hydroxide ion concentrations (≥ 0.01 N).

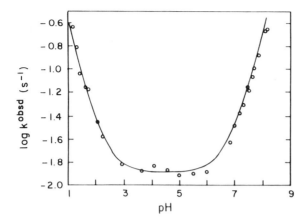

FIGURE 1.   Plot of log k^{obs} vs. pH at [buffer] = 0 for the keto-nization of 3-hydroxy-3,5,7-estratrien-17-one at 25.0° C, μ = 0.1. (Reproduced with permission from Dzingeleski, G. D., Bantia, S., Blotny, G., and Pollack, R. M., *J. Org. Chem.,* 53, 1540, 1988. Copyright 1988, American Chemical Society.)

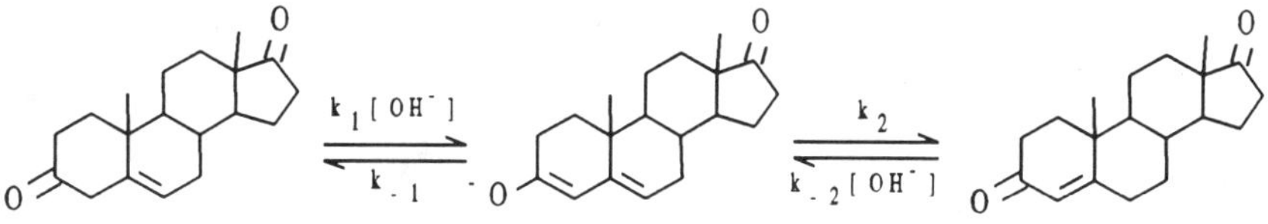

SCHEME 3

This anion may be observed directly in the ultraviolet spectrum when rapid, repetitive scans are taken of **10** in 1.0 N NaOH (Figure 2). Initially, a peak appears with a maximum absorbance of about 257 nm; as the reaction proceeds, this peak is converted smoothly into one characteristic of the product with an absorbance of 248 nm. The rate of formation of the initial species (**12**) was monitored by stopped-flow spectrophotometry at the isosbestic point for the conversion of **12** to **11**. Analysis of the kinetics of the formation of **12** gave a value for the pK_a of **10** of 12.67 ± 0.02, in excellent agreement with the pK_a determined from the overall kinetics of the isomerization.

In order to further characterize the reaction, the exchange rates of the C-6 protons of **11** were determined in deuteromethanol-deuterium oxide solution as a function of DO^-/CH_3O^- concentration. These rate constants, which are measures of k_{-2}, allow the determination of the equilibrium constant for the overall reaction (K = 2400) and the construction of the complete free energy profile for the isomerization. The partitioning of the dienolate ion **12** (k_{-1}/k_2 = 25) shows that protonation of the dienolate occurs preferentially at the

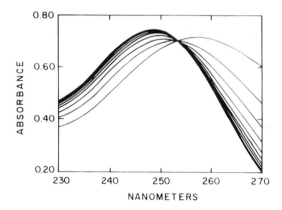

FIGURE 2. Repetitive scans of a solution of 10^{-4} M 5-androstene-3,17-dione in 1.0 M sodium hydroxide. Each scan was taken about 2 to 3 s apart. (Reproduced with permission from Pollack, R. M., Mack, J. P. G., and Eldin, S., *J. Am. Chem. Soc.*, 109, 5048, 1987. Copyright 1987, American Chemical Society.)

α-carbon instead of the γ-carbon, similar to the dienolate ions from both 3-cyclohexenone and 3-cyclopentenone.[8b,16]

The acidities of **10** (pK$_a$ 12.7) and **11** (pK$_a$ 16.1) can be compared with typical saturated ketones such as acetone (pK$_a$ 19.16),[10b] acetophenone (pK$_a$ 18.31),[10c] and isobutyrophenone (pK$_a$ 18.26).[10e] The effect of the β,γ-double bond of **10** on the acidity of the protons at C-4 is about 10^6 to 10^7-fold, whereas the acidifying effect of the α,β-double bond on C-6 protons of **11** is 10^2 to 10^3-fold. A similar acid-strengthening effect of an α-phenyl substituent is seen in the acidity of the benzyl ketones 2-indanone (pK 12.2)[17] and 2-tetralone (pK 12.9).[17a,d]

A dienol intermediate in the isomerization of a β,γ-unsaturated ketone has also been generated in our laboratory by the hydrolysis of the dienol phosphate of 1,3-cyclohexadienol by acid phosphatase.[16] This dienol is the intermediate in the isomerization of 3-cyclohexenone (**13**) to 2-cyclohexenone (**14**) (Equation 6). When the dienol phosphate (**16**) is treated with sweet potato acid phosphatase in slightly acidic aqueous solution (Equation 7), a double exponential decay of absorbance may be observed in the UV spectrum at 265 nm (Figure 3). Analysis of the kinetics produces rate constants for both the formation of the intermediate dienol and its decomposition to **13**. The final product gives virtually no absorbance in the UV spectrum, demonstrating that partitioning of **15** under these conditions ($4.2 <$ pH < 5.2, acetate buffer) gives almost exclusively 3-cyclohexenone. An estimate of the partitioning ratio ($k_{-1}/k_2 \approx 45$) was obtained by comparing the rate of exchange of the protons at C-2 in acetate buffer (D_2O) with the overall rate of isomerization in acetate buffer (H_2O).[16b] Since the overall equilibrium constant has previously

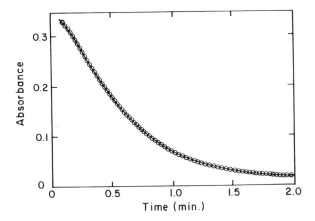

FIGURE 3. Plot of absorbance vs. time for the reaction of sweet potato acid phosphatase (0.0375 mg/ml) with 1,3-cyclohexadienol phosphate in deuterium oxide solution (pD 4.70, [OAc$^-$] = 5 mM, 25°C). (Reproduced with permission from Pollack, R. M., Mack, J. P. G., and Blotny, G., *J. Am. Chem. Soc.,* 109, 3138, 1987. Copyright 1987, American Chemical Society.)

been measured,[8b] there is sufficient information to construct the complete free energy profile for the general base-catalyzed (acetate) isomerization of **13**, allowing the keto-enol equilibrium constants to be determined for both **13** ($K_E = k_1/k_{-1} = 4.4 \times 10^{-6}$) and **14** ($K_E = k_{-2}/k_2 = 1.5 \times 10^{-8}$).

$$
\underset{13}{\text{[structure]}} \underset{k_{-1}}{\overset{k_1}{\rightleftharpoons}} \underset{15}{\text{[structure]}} \underset{k_{-2}}{\overset{k_2}{\rightleftharpoons}} \underset{14}{\text{[structure]}} \tag{6}
$$

$$
\underset{16}{\text{[structure]}} \overset{\text{P-ase}}{\longrightarrow} \underset{15}{\text{[structure]}} \longrightarrow \underset{13}{\text{[structure]}} \tag{7}
$$

IV. INTERMEDIATES IN THE STEROID ISOMERASE REACTION

The enzyme steroid isomerase (KSI) from *P. testosteroni* has been the object of intense investigation during the last 35 years.[18] The isomerase catalyzes the conversion of 5-androstene-3,17-dione (**10**) to 4-androstene-3,17-dione (**11**). This activity was first observed in 1955, and since that time the enzyme has been crystallized[19] and sequenced.[20] In addition, X-ray crystallographic investigations have yielded structures with resolution to 2.5 Å;[21] the enzyme has been cloned and overexpressed,[22] and site-directed mutagenesis on the active site has been performed.[23] The impetus for investigations of the isomerase comes from (1) the role of the mammalian equivalent in the biosynthesis of all classes of steroid hormones, (2) its high activity ($k_{cat} = 4 \times 10^6$ min^{-1}), and (3) the simplicity of the catalytic reaction.

A priori, an attractive mechanism for the action of the isomerase is the formation of a Schiff base intermediate from an enzymatic lysine residue and the carbonyl at the 3-position, by analogy with the mechanism of isomerization of 3-methyl-3-cyclohexenone catalyzed by primary amines. However, several lines of evidence argue against this mechanism: (1) incubation of the non-covalent complex of radioactively labeled 19-nortestosterone (a competitive inhibitor) and the enzyme with sodium borohydride results in neither the loss of activity nor the incorporation of radioactivity into the enzyme;[24] similarly, (2) an attempt to inactivate the isomerase by reduction with borohydride in the presence of the substrate 5-androstene-3,17-dione was unsuccessful;[25] (3) no inhibition was found with cyanide ion,[15] which would be expected to add to a Schiff base intermediate,[25] and finally, (4) modification of the amino groups of the isomerase with methyl acetimidate resulted in no loss of activity.[26]

The currently accepted mechanism involves a single base acting as a proton shuttle to transfer a hydrogen from the C-4β to the C-6β position, through either a dienol or a dienolate ion intermediate (Scheme 4). Site-directed mutagenesis studies suggest a role for Tyr-14 and Asp-38, probably as a proton donor (or hydrogen bonding group) to the carbonyl and the base, respectively.[23]

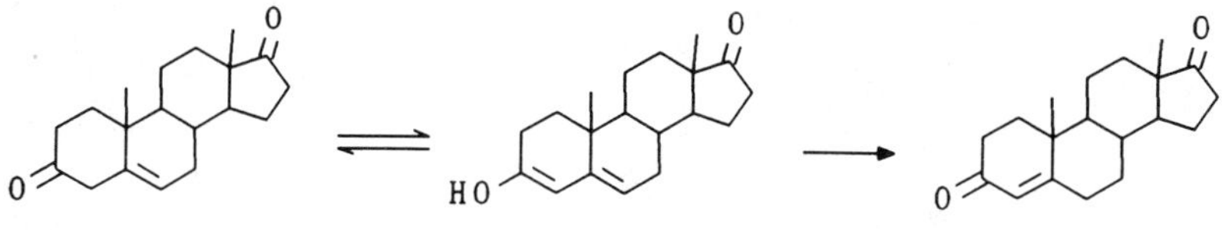

SCHEME 4

In order to evaluate the possible intermediacy of a dienol(ate), we examined[27] the isomerase-catalyzed conversion of 5,7-estradiene-3,17-dione (**7**) to 4,7-estradiene-3,17-dione (**8**) (Equation 8). Compound **7** has an extra double bond (Δ^7) and one less methyl group (at C-19) than the usual substrate 5-androstene-3,17-dione (**10**). However, **7** is an excellent substrate, producing **8** at a rate only slightly slower than the corresponding reaction with **10**. Interestingly, none of the fully conjugated ketone is formed. Since the putative intermediate trienol (**6**) in this reaction is isolable, we were able to examine its reaction with the isomerase. The trienol is an excellent substrate for the isomerase, being converted to the product at a rate comparable to the overall rate of isomerization of **7** → **8** (Equation 9). Thus, the dienol (or dienolate) is implicated in the mechanism.

$$(8)$$

$$(9)$$

More recently,[28] we have been able to generate the dienol intermediate for the isomerization of 5-androstene-3,17-dione (**10**) and determine its partitioning at the active site of KSI. When the dienol phosphate of **17** is incubated with sweet potato acid phosphatase in acetate buffer (pH 5.1), cleavage of the oxygen-phosphorous bond leads to formation of the dienol (**17**) in solution. Greater than 99% of this dienol can be trapped by 0.01 to 0.03 μM KSI and converted to a mixture of **10** and **11**. When the reaction is run in the presence of α-hydroxysteroid dehydrogenase (HSD) and NADH, the **10** that is formed is reduced before further reaction with KSI produces **11** (Scheme 5). Extrapolation of the product ratio [**18**]/[**11**] to infinite [HSD] shows that, under these conditions, about 65% of the intermediate dienol is converted to the unconjugated ketone (**10**) and 35% to the conjugated ketone (**11**) by KSI.

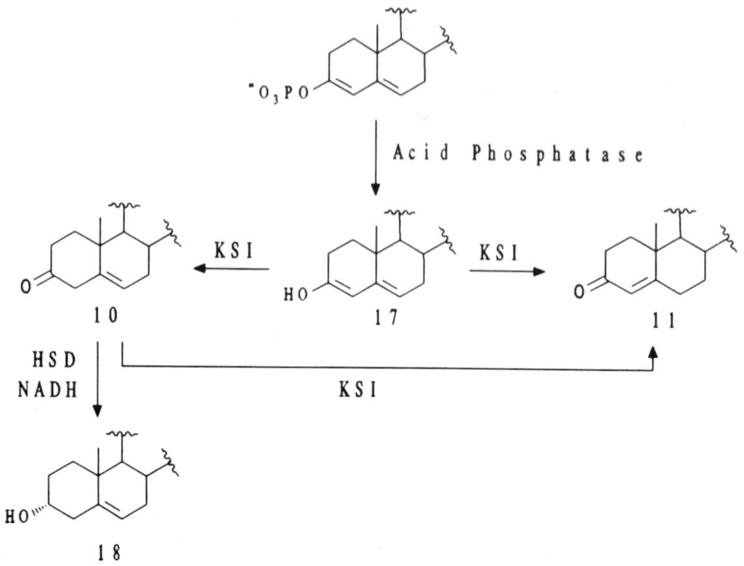

SCHEME 5

We were also able to monitor the rate of reaction of the dienol with KSI by stopped-flow spectrophotometry. The dienol (**17**) was generated by rapidly quenching a solution of the dienolate ion (**12**, formed by mixing equal volumes of **10** and 1.0 *N* NaOH) into buffer solution. Since **10** is quite acidic for a ketone (pK_a 12.7), the anion is formed in large amounts in strongly basic aqueous solutions. Quenching into buffer solutions (acetate or phosphate) leads to protonation on oxygen to generate the dienol (**17**). When **17** is prepared in the presence of KSI, its conversion to a mixture of **10** and **11** can be monitored by observing the change in absorbance at the isosbestic point of **17** to **11** (Scheme 6 and Figure 4). The initial loss of absorbance corresponds to conversion of **17** to **10**, whereas the subsequent rise in absorbance is due to conversion of **10** to **11** by KSI. Analysis of the absorbance variation with time, coupled with the extent of loss of absorbance at the minimum, allows the determination of the rate constants for the conversion of **17** to **10** (k_a) and to **11** (k_b). These values, along with the rate constant for enzymatic conversion of **10** to **11**, are listed in Table 3.

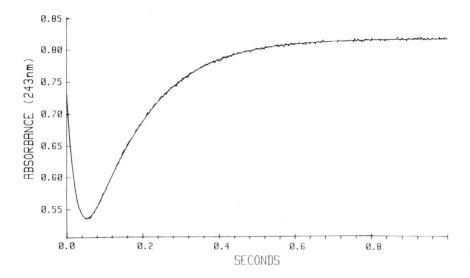

FIGURE 4. Absorbance change at 243 nm for the reaction of dienol **10** with KSI. Solutions of **10** (5 × 10⁻⁴ *M* in 20% MeOH) and 1.0 *N* NaOH were mixed in a 1:1 ratio and allowed to age for about 0.5 s. This solution was then rapidly quenched with five volumes of a pH 6.8 phosphate solution containing KSI (400 m*M* phosphate, 2.0% MeOH, 0.050 μ*M* KSI) in the observation chamber of a stopped-flow spectrophotometer (final pH 7.3, 3.3% MeOH, 25.0°C). The theoretical line is calculated using k_a = 9.6 s⁻¹, k_b = 7.7 s⁻¹, and k_c = 11 s⁻¹.

TABLE 3
Rate Constants for the Reaction of Dienol 10 with KSI (3.3% Methanol, 25°C)

pH	$10^{-8} k_a$ $(M^{-1} s^{-1})$	$10^{-8} k_b$ $(M^{-1} s^{-1})$	$10^{-8} k_c$ $(M^{-1} s^{-1})$	$k_c/(k_c + k_a)$ (%)
5.1	1.4 ± 0.1	0.37 ± 0.04	0.65 ± 0.12	31.4 ± 3.1
7.0	2.1 ± 0.2	2.1 ± 0.4	2.3 ± 0.3	53.2 ± 3.1
7.7	2.5 ± 0.3	2.5 ± 0.3	3.1 ± 0.4	55.4 ± 2.0

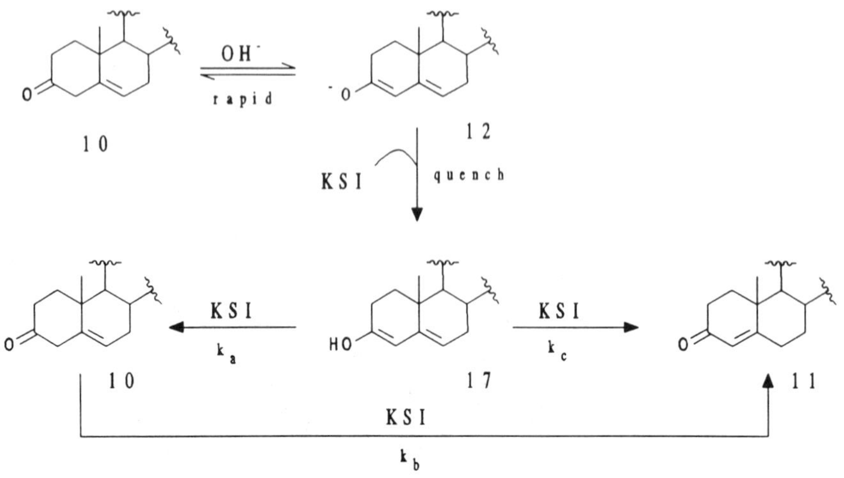

SCHEME 6

A comparison of the overall rate constant for the reaction of the dienol with KSI with the corresponding value for the Δ^5-ketone shows that the dienol is kinetically competent, consistent with its role as an intermediate in the enzymatic conversion of **10** to **11**. The nearly 1:1 partitioning ratio of the dienol when it reacts with KSI suggests that the barriers to protonation at C-4 (to produce **10**) and at C-6 (to produce **11**) are similar. This result appears to be the first example of the direct determination of the partitioning of an active intermediate by an enzyme.

ACKNOWLEDGMENT

The author would like to acknowledge both the intellectual stimulation and encouragement that he received from Myron Bender as a member of his laboratory during the period 1968 to 1970, as well as in later years. Financial support, of course, had to be obtained from other sources, and the author thanks the National Institutes of Health, the American Cancer Society, and the Petroleum Research Fund of the American Chemical Society for the support of the research from his laboratory. Finally, the intellectual and experimental contributions of the members of the author's own research group made this work possible.

REFERENCES

1. (a) **Bender, M. L.,** *J. Am. Chem. Soc.,* 83, 1626, 1951; (b) **Bender, M. L., Ladenheim, H., and Chen, M. C.,** *J. Am. Chem. Soc.,* 83, 123, 1961; (c) **Bender, M. L. and Thomas, R.,** *J. Am. Chem. Soc.,* 83, 4183, 1961; (d) **Bender, M. L. and Thomas, R.,** *J. Am. Chem. Soc.,* 83, 4189, 1961; (e) **Bender, M. L., Matsui, H., Thomas R., and Tobey, S. W.,** *J. Am. Chem. Soc.,* 83, 4199, 1961; (f) **Bender, M. L. and Heck, H. d'A.,** *J. Am. Chem. Soc.,* 89, 1211, 1967.

2. (a) **Schonbaum, G. R., Nakamura, K., and Bender, M. L.,** *J. Am. Chem. Soc.,* 81, 4746, 1959; (b) **Bender, M. L. and Zerner, B.,** *J. Am. Chem. Soc.,* 83, 2391, 1961; (c) **Bender, M. L., Kaiser, E. T., and Zerner, B.,** *J. Am. Chem. Soc.,* 83, 4656, 1961; (d) **Bender, M. L., Schonbaum, G. R., and Zerner, B.,** *J. Am. Chem. Soc.,* 84, 2540, 1962; (e) **Bender, M. L. and Zerner, B.,** *J. Am. Chem. Soc.,* 84, 2550, 1962; (f) **Bender, M. L. and Kaiser, E. T.,** *J. Am. Chem. Soc.,* 84, 2556, 1962; (g) **Bender, M. L. and Clement, G. E.,** *Biochem. Biophys. Res. Commun.,* 12, 339, 1963; (h) **Kezdy, F. J. and Bender, M. L.,** *J. Am. Chem. Soc.,* 86, 937, 1964; (i) **Zerner, B., Bond, R. P. M., and Bender, M. L.,** *J. Am. Chem. Soc.,* 86, 3674, 1964; (j) **Kezdy, F. J., Clement, G. E., and Bender, M. L.,** *J. Am. Chem. Soc.,* 86, 3690, 1964; (k) **Bender, M. L. and Brubacher, L. J.,** *J. Am. Chem. Soc.,* 86, 5333, 1964; (l) **Miller, C. G. and Bender, M. L.,** *J. Am. Chem. Soc.,* 90, 6850, 1968.

3. For a review of the isomerization reaction: **Pollack, R. M., Bevins, C. L., and Bounds, P. L.,** in *The Chemistry of the Functional Groups, Enones,* Patai, S. and Rappoport, Z., Eds., John Wiley & Sons, Chichester, 1989.

4. **Bender, M. L. and Williams, A.,** *J. Am. Chem. Soc.,* 88, 2502, 1966.

5. (a) **Hine, J., Menon, B. C., Jensen, J. H., and Mulders, J.,** *J. Am. Chem. Soc.,* 88, 3367, 1966; (b) **Hine, J., Kokesh, F. C., Hampton, K. G., and Mulders, J.,** *J. Am. Chem. Soc.,* 89, 1205, 1967; (c) **Hine, J., Mulders, J., Houston, J. G., and Idoux, J. P.,** *J. Org. Chem.,* 32, 2205, 1967.

6. (a) **Kayser, R. H. and Pollack, R. M.,** *J. Am. Chem. Soc.,* 97, 952, 1975; (b) **Pollack, R. M. and Kayser, R. H.,** *J. Am. Chem. Soc.,* 98, 4174, 1976.

7. (a) **Brault, M. and Pollack, R. M.,** *J. Org. Chem.,* 41, 346, 1976; (b) **Brault, M. and Pollack, R. M.,** *J. Am. Chem. Soc.,* 98, 247, 1976; (c) **Brault, M., Kayser, R. H., and Pollack, R. M.,** *J. Org. Chem.,* 43, 4709, 1978.

8. (a) **Nes, W. R., Loesser, E., Kirdant, R., and Marshy, J.,** *Tetrahedron,* 19, 299, 1963; (b) **Whalen, D. L., Weimaster, J. F., Ross, A. M., and Radhe, R.,** *J. Am. Chem. Soc.,* 98, 7319, 1976; (c) **Westphal, U. and Schmidt-Thome, J.,** *Chem. Ber.,* 69, 889, 1936; (d) **Malhotra, S. K. and Ringold, H. J.,** *J. Am. Chem. Soc.,* 87, 3228, 1965; (e) **Noyce, D. S. and Evett, M.,** *J. Org. Chem.,* 37, 394, 397, 1972; (f) **Jones, J. B. and Wigfield, D. C.,** *Can. J. Chem.,* 47, 4459, 1969; (g) **Kergomard, A., Xang, L. Q., and Renard, M. F.,** *Tetrahedron,* 32, 1989, 1976.

9. (a) **Okuyama, T., Kitada, A., and Fueno, T.,** *Bull Chem. Soc. Jpn.,* 50, 2358, 1977; (b) **Perera, S. K., Dunn, W. A., and Fedor, L. R.,** *J. Org. Chem.,* 45, 2816, 1980.

10. (a) **Chiang, Y., Kresge, A. J., and Walsh, P. A.,** *J. Am. Chem. Soc.,* 104, 6122, 1982; (b) **Chiang, Y., Kresge, A. J., Tang, Y. S., and Wirz, J.,** *J. Am. Chem. Soc.,* 106, 460, 1984; (c) **Chiang, Y., Kresge, A. J., and Wirz, J.,** *J. Am. Chem. Soc.,* 106, 6392, 1984; (d) **Chiang, Y., Kresge, A. J., and Walsh, P. A.,** *J. Am. Chem. Soc.,* 108, 6314, 1986; (e) **Pruszynski, P., Chiang, Y., Kresge, A. J., Schepp, N. P., and Walsh, P. A.,** *J. Phys. Chem.,* 90, 3760, 1986; (f) **Kresge, A. J.,** *Chem. Tech.,* 16, 250, 1986; (g) **Keeffe, J. R., Kresge, A. J., and Schepp, N. P.,** *J. Am. Chem. Soc.,* 110, 1993, 1988; (h) **Kresge, A. J.,** *Acc. Chem. Res.,* 23, 43, 1990; (i) **Keeffe, J. R., Kresge, A. J., and Schepp, N. P.,** *J. Am. Chem. Soc.,* 112, 4862, 1990.

11. (a) **Capon, B., Rycroft, D. S., and Watson, T. W.,** *J. Chem. Soc. Chem. Commun.,* 724, 1979; (b) **Capon, B., Rycroft, D. S., Watson, T. W., and Zucco, C.,** *J. Am. Chem. Soc.,* 103, 1761, 1981; (c) **Croft, B. and Zucco, C.,** *J. Am. Chem. Soc.,* 104, 7567, 1982; (d) **Capon, B. and Siddhanta, A. K.,** *J. Org. Chem.,* 49, 255, 1984; (e) **Capon, B., Siddhanta, A. K., and Zucco, C.,** *J. Org. Chem.,* 50, 3580, 1985; (f) **Capon, B., Guo, B.-Z., Kwok, F. C., Siddhanta, A. K. and Zucco, C.,** *Acc. Chem. Res.,* 21, 135, 1988.

12. (a) **Noyori, R., Inoue, H., and Katô, M. J.,** *J. Am. Chem. Soc.,* 92, 6699, 1970; (b) **Noyori, R., Inoue, H., and Katô, M. J.,** *Bull. Chem. Soc. Jpn.,* 49, 3673, 1976; (d) **Duhaime, R. M. and Weedon, A. C.,** *J. Am. Chem. Soc.,* 107, 6723, 1985; (e) **Duhaime, R. M. and Weedon, A. C.,** *J. Am. Chem. Soc.,* 109, 2479, 1987; (f) **Duhaime, R. M. and Weedon, A. C.,** *Can. J. Chem.,* 65, 1867, 1967; (g) **Capon, B.,** in *The Chemistry of the Functional Groups, Enones,* Patai, S. and Rappoport, Z., Ed., John Wiley & Sons, Chichester, 1989.

13. **Kruger, G. J.,** *Org. Chem.,* 33, 1750, 1968.

14. **Dzingeleski, G. D., Bantia, S., Blotny, G., and Pollack, R. M.,** *J. Org. Chem.,* 53, 1540, 1988.

15. (a) **Pollack, R. M., Mack, J. P. G., and Eldin, S.,** *J. Am. Chem. Soc.,* 109, 5048, 1987; (b) **Pollack, R. M., Zeng, B., Mack, J. P. G., and Eldin, S.,** *J. Am. Chem. Soc.,* 111, 6419, 1989.

16. (a) **Pollack, R. M., Mack, J. P. G., and Blotny, G.,** *J. Am. Chem. Soc.,* 109, 3138, 1987; (b) **Dzingeleski, G., Blotny, G., and Pollack, R. M.,** *J. Org. Chem.,* 55, 1019, 1990.

17. (a) **Ross, A. M., Whalen, D. L., Eldin, S., and Pollack, R. M.,** *J. Am. Chem. Soc.,* 110, 1981, 1988; (b) **Keeffe, J. R., Kresge, A. J., and Lin, Y.,** *J. Am. Chem. Soc.,* 110, 1982, 1988; (c) **Keeffe, J. R., Kresge, A. J., and Lin, Y.,** *J. Am. Chem. Soc.,* 110, 8201, 1988; (d) **Eldin, S., Pollack, R. M., and Whalen, D. L.,** *J. Am. Chem. Soc.,* 113, 1344, 1991.

18. For a recent review of the mechanism of action of steroid isomerase, see Reference 3.

19. **Kawahara, F. S. and Talalay, P.,** *J. Biol. Chem.,* 235, PC1, 1960.

20. **Benson, A. M., Jarabak, R., and Talalay, P.,** *J. Biol. Chem.,* 246, 7514, 1971.

21. **Westbrook, E. M.,** personal communication.

22. (a) **Kuliopulos, A., Shortle, D., and Talalay, P.,** *Proc. Natl. Acad. Sci. U.S.A.,* 84, 8893, 1987; (b) **Choi, K. Y. and Benisek, W. F.,** *Gene,* 58, 257, 1987.

23. (a) **Kuliopulos, A., Mildvan, A. S., Shortle, D., and Talalay, P.,** *Biochemistry,* 28, 149, 1989; (b) **Xue, L., Talalay, P., and Mildvan, A. S.,** *Biochemistry,* 29, 7491, 1990; (c) **Kuliopulos, A., Talalay, P., and Mildvan, A. S.,** *Biochemistry,* 29, 10271, 1990.

24. **Batzold, F. H., Benson, A. M., Covey, D. F., Robinson, C. H., and Talalay, P.,** *Adv. Enzyme Regul.,* 14, 243, 1976.

25. **Pollack, R. M.,** unpublished results.

26. **Benisek, W. F. and Jacobson, A.,** *Bioorg. Chem.,* 4, 41, 1975.

27. **Bantia, S. and Pollack, R. M.,** *J. Am. Chem. Soc.,* 108, 3145, 1986.

28. (a) **Eames, T. C. M., Hawkinson, D. C., and Pollack, R. M.,** *J. Am. Chem. Soc.,* 112, 1996, 1990; (b) **Hawkinson, D. C., Eames, T. C., and Pollack, R. M.,** *Biochemistry,* 30, 6956, 1991.

Chapter 10

BORONIC ACIDS CATALYZE THE HYDROLYSIS OF MANDELONITRILE

Galla Rao and Manfred Philipp

TABLE OF CONTENTS

I. INTRODUCTION

The chemistry of boric and boronic acids has long been of interest to chemists due to the Lewis acid character of these compounds. Boron in trivalent compounds has one vacant orbital which is available for the formation of a fourth covalent bond with electron donor atoms. Boric and boronic acids are trigonal compounds and ionize to form tetrahedral anions by accepting an electron pair from OH^- to form tetrahedral $R\text{-}B(OH)_3^-$.[1]

In the 1870s, it was observed that the acidity of boric acid solutions is increased by the addition of glycerol.[2] In the early part of this century, several investigators studied the formation of boric acid complexes with polyols (carbohydrates), phenols, cyclic glycols, and hydroxy acids.[2] The extent of complex formation was determined by measuring the enhancement of boric acid solution conductivity on addition of alcohols. Boeseken[2] applied this method to determine the configuration of carbohydrates. Boric and boronic acids also form esters with simple alcohols; the esters of unhindered primary alcohols undergo hydrolysis very rapidly in water to regenerate the boron acids.[3]

In recent years, Pizer and colleagues[4-9] have studied the reaction of boric acid, benzeneboronic acid, and substituted benzeneboronic acids with polyols, dicarboxylic acids, and hydroxy acids using temperature-jump relaxation kinetics. They determined rate and equilibrium constants for complex formation. The dissociation constants are in the millimolar to micromolar range, and depend upon the acidities of the boronic acids and the ligands.

The ability of boric and boronic acids to form reversible complexes with various ligands has had several applications. One is in affinity chromatography, where a boronic acid can be attached to cellulose derivatives and then be used to separate diols.[10]

The second application is the use of boronic acids as transition state analog inhibitors for serine proteases. In 1957, Torssell[11] noticed the inhibition of serum cholinesterase by benzeneboronic acid. Philipp and Bender[12] showed that substituted benzeneboronic acids inhibit chymotrypsin and subtilisin. Peptideboronic acids are also used as inhibitors and are found to inhibit enzymes in the nanomolar concentration range.[13,14] Koehler and Lienhard[15] proposed that boronic acids are transition state analogs. Boronic acids esterify to the serine hydroxyl group at the active site of these enzymes. As a result, the configuration about the boron atom becomes tetrahedral and mimics the tetrahedral transition state of the enzyme-catalyzed ester hydrolysis.

The third application of boronic acids has been their use as enzyme models. The ability of boronates to reversibly complex to alcohols in aqueous solution kinetically mimics the formation of an enzyme-substrate complex.

In 1960, Peer[16] found that phenol, in the presence of boric acid, reacts with formaldehyde and gives *o*-hydroxymethylphenol as the only product. This product does not form in the absence of boric acid. The specificity of the reaction is probably due to the rapid and reversible formation of a complex of phenol, borate, and formaldehyde (Equation 1).

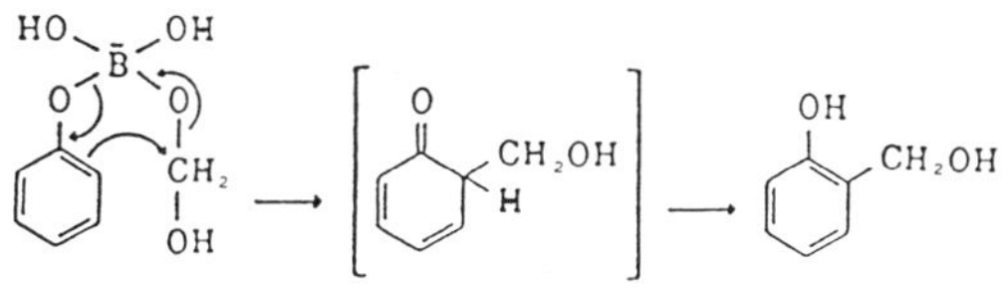

(1)

(From Jencks, W. P., *Catalysis in Chemistry and Enzymology,* McGraw-Hill, New York, 1969, 30. With permission.)

In 1963, Letsinger et al.[18] used 8-quinolineboronic acid as a catalyst in the hydrolysis of chloroethanol to ethyleneglycol. Quinolineboronic acid catalyzes the hydrolysis at least 80 times faster than a mixture of quinoline and benzeneboronic acid, showing the advantage of intramolecular catalysis over intermolecular catalysis. In this reaction, the boronic acid residue esterifies to the alcohol group of the substrate prior to reaction of the quinoline nitrogen atom with the halogenated alcohol.

8-Quinolineboronic acid also shows *cis-trans* stereoselectivity in the hydrolysis of chloroalcohols.[19] In the case of 2-chloro-1-indanol, it preferentially hydrolyzes the *trans* isomer.

In another study, Letsinger and Macheon[20] used boronoarylbenzimidazole as a catalyst in the formation of ethers from chloroethanol in butanol solution. Here also, the borono group in boronoarylbenzimidazole binds the alcohol substrates and holds them in a position favorable for reaction.

In 1966, Capon and Ghosh[21] found that borate catalyzes the hydrolysis of phenyl salicylate more than 100-fold more rapidly than the hydrolysis of phenyl-*o*-methoxybenzoate and phenyl benzoate. Borate apparently first forms a complex with the ester, the boron atom then acting as a Lewis acid to accept a lone pair of electrons from the carbonyl oxygen atom, leading to the formation of an intermediate. This intermediate resembles the transition state of ester hydrolysis (Equation 2).

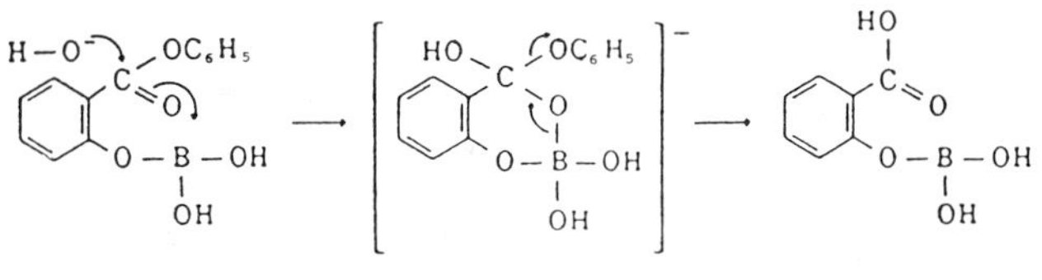

(2)

(From Jencks, W. P., *Catalysis in Chemistry and Enzymology,* McGraw-Hill, New York, 1969, 30. With permission.)

Okuyama et al.[22] showed that borate catalyzes the hydrolysis of S-butyl 2-hydroxy-2-phenylthioacetates by a factor of about 80 at pH 9. Butylthioacetate, which has no α-hydroxyl group, is not hydrolyzed by borate.

We have found that boronic acids catalyze the hydrolysis of salicylaldehyde imines. This catalysis shows pronounced pH dependencies, with activity depending on the trigonal neutral boronic acid. The reactions show clear Michaelis-Menten kinetics, a positive Hammett ρ for binding of substrate to catalyst, and a nearly zero (-0.064) value of ρ for the catalytic step. The mechanism of the catalysis may be an intramolecular transfer of a boron-coordinated hydroxide ion within a borate-substrate complex (Equation 3[23]):

$$
\begin{array}{c}
\text{(salicylaldehyde imine substrate)} + R\text{-B(OH)}_2 \;\rightleftharpoons\; \text{(salicylaldehyde, CHO/OH)} + \text{(amino acid fragment)} + B
\end{array}
\tag{3}
$$

Boric acids have also been shown to catalyze the formation and hydrolysis of hydroxy[24] and salicylaldehyde imines.[25,27] In the studies done with imines, boronic and borinic acids were found to be better catalysts than boric acid,[23] mainly due to kinetic effects resulting from boron acid pK values on the binding component of overall rate constants.

Nitriles and imines have obvious structural similarities. Both the nitriles and the imines have unsaturated bonds connecting carbon and nitrogen atoms; both can be hydrolyzed to form a carbonyl compound. Given our results showing that boronic acids can catalyze the hydrolysis of hydroxy-containing imines, the obvious question was if these acids can catalyze the hydrolysis of hydroxyl group-containing nitriles. The patent literature[28] shows work describing the boric acid-catalyzed mandelonitrile hydrolysis to form mandelamide. In the expectation that the mechanism of this reaction might resemble that of the imines, we decided to investigate the mechanism of nitrile hydrolysis using substituted boronic acids.

II. EXPERIMENTAL

A. MATERIALS

Deuterium oxide purchased from Aldrich Buffers was prepared in D_2O. Initial 0.5 M NaOH and 1.0 M HCl solutions were prepared by dissolving alkali or by diluting concentrated acid in a small volume of D_2O; 0.5 M KH_2PO_4 and 1.0 M $NaHCO_3$ solutions were prepared in a similar way. Phosphate buffers and bicarbonate buffers of 0.1 M were prepared with the above solutions by using buffer formulas given in the *Biochemists' Handbook*.[29]

Mandelamide was purchased from ICN Pharmaceuticals 3-(Trimethylsilyl)-1-propanesulfonic acid was purchased from Aldrich. Dimethylsulfoxide-d_6, and tetramethylsilane were purchased from Norell.

Benzeneboronic acid, 3-aminobenzeneboronic acid, and 4-bromobenzene-boronic acid were purchased from Aldrich. 3-Nitrobenzeneboronic acid was purchased from ICN, 3-carboxybenzeneboronic acid was purchased from Calbiochem, 3,5-*bis*-(trifluoromethyl)benzeneboronic acid was purchased from Alpha, and 4-toluene boronic acid was prepared according to the method of Bean and Johnson.[30] Buffers containing the appropriate concentrations of boronic acid were prepared by dissolving the boronic acid in buffer and adjusting the pH with NaOH or CHl.

D,L-Mandelonitrile was prepared by the procedure of Jorns.[31] Product identity was determined by observation of the NMR spectrum (Figure 1). A 0.83 M stock solution was made in deuterated dimethylsulfoxide.

B. METHODS

The hydrolysis of mandelonitrile was carried out by following the appearance of a specific proton peak of the product at 5.15 ppm using a computer-interfaced JEOL GX 400 NMR Spectrometer. The spectrometer was equipped with a variable temperature controller which maintains a constant temperature within $+0.5°C$.

One ml of buffer containing the appropriate concentration of boronic acid was mixed with 0.1 ml of mandelonitrile from a stock solution. It was then transferred to a 5-mm NMR tube (Wilmad 528 pp) and frozen until the start of the experiment. 3-(Trimethylsilyl)-1-propanesulfonic acid was used as an internal standard. The reaction mixture was brought to room temperature before the experiment and then maintained at 60°C in the ${}^1H$ probe. The NMR spectra of the reaction mixture were recorded at 6-min intervals for 90 min and stored on a disk. After 90 min, spectra were printed with the integration value of several protons. The time-dependent integrated areas of each peak were measured and used in the calculation of first-order rate constants. The area of appropriate peaks at infinite time were used in the first-order calculations.

C. RESULTS

As shown in Equation 4, mandelonitrile (**1**), in the presence of boronic acids (**B**), undergoes hydrolysis to mandelamide (**2**). The NMR spectrum of the reaction product agrees with the spectrum of mandelamide (Figure 1). The identity of the product was further confirmed by TLC on a silica gel (Silica gel 60 F_{254}, MC/B, solvent system: hexane:butanol, 50:50) and by comparing the R_f value (0.35) of the product with that of an authentic sample. No such reaction could be observed when using benzonitrile (**1A**) instead of mandelonitrile.

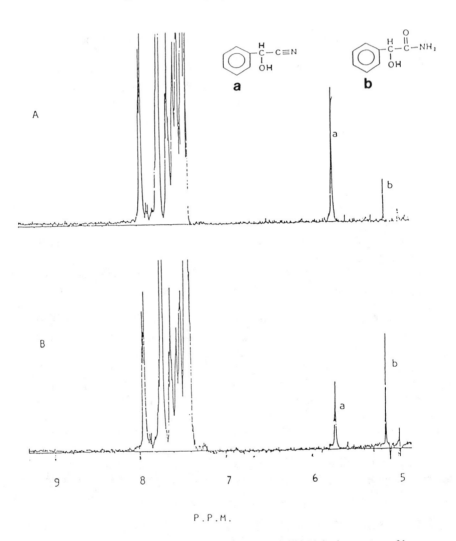

FIGURE 1.   400-MHz NMR spectra of mandelonitrile (0.076 *M*) in the presence of benzene-boronic acid (0.05 *M*) in 0.1 *M* bicarbonate buffer, D$_2$O at pH 9.0 and 60°C. (A) At zero time; (B) after 78 min. Peaks are from mandelonitrile (a) and mandelamide (b).

In 1978, Wechsberg and Schönbeck showed that mandelamide can be prepared by reacting mandelonitrile with aqueous borate solutions at 60°C.[28] Since they reported no kinetic studies, the mechanism of the reaction is unknown. Our preliminary studies showed that the elevated temperature is necessary; the rate of hydrolysis is very slow at room temperature. In control experiments, we found that both the reactant and product were not affected by high temperature in the absence of boron acids. There were no differences in the spectra of reactants and products at 25 and 60°C. Spontaneous nitrile or amide hydrolysis was undetectable under these conditions. All of the studies described below were done at 60°C.

1. Effect of Benzeneboronic Acid Concentration on Mandelonitrile Hydrolysis

The rate of hydrolysis of mandelonitrile was studied by varying the concentration of benzeneboronic acid in 0.1 M bicarbonate buffer, pH 9.0. The rate of hydrolysis increased continuously with increasing concentration of boronic acid. The highest concentration of boronic acid used was 0.3 M. These data show no evidence of mandelonitrile-boronate complexation at this concentration.

2. Effect of pH on the Hydrolysis of Mandelonitrile by Boronic Acids

The rate of benzeneboronic acid-catalyzed hydrolysis of mandelonitrile was studied in the pH range 7.5 to 10.4 (Figure 2). There was no spontaneous hydrolysis in this pH range and no detectable amount of product formation below pH 7.5. The rate continuously increases with increasing pH. This is opposite to that seen in the imine hydrolysis by boronic acid.[23] When the logarithm of first-order rate constants are plotted as a function of pH, a straight line is obtained with a slope of -0.4 (see Figure 4).

Figure 3 shows the pH dependence of mandelonitrile hydrolysis by 3-nitrobenzeneboronic acid. This was done in order to determine if different boronic acid pK values result in different catalytic pH profiles. There was no detectable amount of product formation below pH 8.0 and the rate continuously increased with increasing pH, similar to that of benzeneboronic acid. A straight line with a slope of -0.17 is obtained when log first-order rate constants are plotted as a function of pH (Figure 4).

3. Effect of Substituents of Benzeneboronic Acid on the Hydrolysis of Mandelonitrile

This experiment was done in order to determine the influence of Hammett σ on catalytic rate constants. Catalytic rates for substituted benzeneboronic acids were studied in 0.1 M bicarbonate buffer, pH 9.2, at 60°C. Observed first-order rate constants were converted to second-order rate constants by dividing by the boronic acid concentration. Boronic acids with electron-donating substituents have higher second-order rate constants.

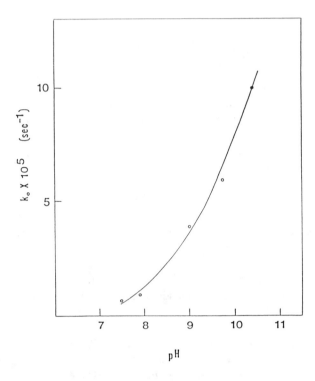

FIGURE 2.   pH profile for the hydrolysis of mandelonitrile (0.076 *M*) by benzeneboronic acid (0.05 *M*) at 60°C. Buffers below pH 8 are 0.1 *M* phosphate and above pH 9 are 0.1 *M* bicarbonate. The hydrolysis was followed by integrating the proton peak of the product at 5.15 ppm with time (every 6 min), using a 400-MHz NMR spectrometer.

Figure 5 shows a Hammett plot of second-order rate constants for the hydrolysis of mandelonitrile by boronic acids vs. substituent constants. The observed value of Hammett ρ is -0.75.

III. DISCUSSION

In the presence of boronic acids (**B**), mandelonitrile (**1**) undergoes hydrolysis to mandelamide (**2**). Benzonitrile (**1A**), which lacks a hydroxyl group adjacent to the nitrile, is not affected by the presence of boronic acids. Since boronic acids reversibly esterify to alcohols, this suggests that the hydrolysis of mandelonitrile depends on the formation of a complex between boronic acids and mandelonitrile, a complex dependent on the mandelonitrile hydroxyl group.

Precedents for the formation of such a complex include the boronic acid-catalyzed hydrolysis of imines[23] and the boric acid-catalyzed hydrolysis of α-hydroxythioesters.[22]

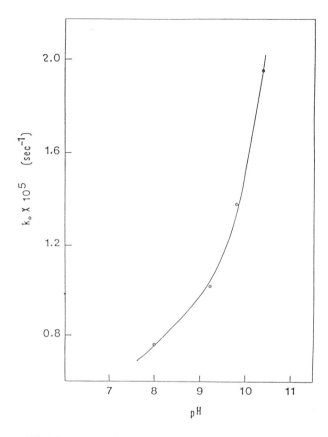

FIGURE 3. pH profile for the hydrolysis of mandelonitrile (0.076 *M*) by 3-nitrobenzeneboronic acid (0.05 *M*) at 60°C. Buffers below pH 8 are 0.1 *M* phosphate and above pH 9 are 0.1 *M* bicarbonate. The hydrolysis was followed by integrating the proton peak of the product at 5.15 ppm with time (every 6 min), using a 400-MHz NMR spectrometer.

If the boronic acid forms a tightly binding complex with mandelonitrile, then a hyperbolic curve might be expected when catalytic rate constants are plotted as a function of boronic acid concentration. However, plots relating boronic acid concentration and reaction rate show that the relationship is linear even at the highest boronic acid concentrations.

This linearity may be due to a very high dissociation constant for the presumed benzeneboronic acid-mandelonitrile complex. Such a high value may be expected for a monodentate ligand.[32]

The hydrolysis of mandelonitrile by benzeneboronic acid is pH dependent and increases with increasing pH (Figure 2). The pH profiles are very different from those observed in the hydrolysis of an imine by boronic acids,[23] where the hydrolysis decreases with increasing pH. It is not possible to determine

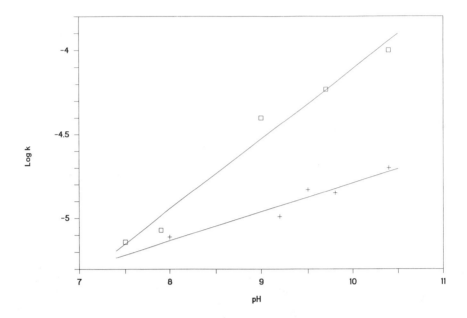

FIGURE 4.   Plot of log first-order rate constant for the hydrolysis of mandelonitrile (0.076 *M*) by benzeneboronic acid (□) and 3-nitrobenzeneboronic acid (+) as a function of pH. The slopes of the plots are + 0.4 (benzeneboronic acid) and + 0.17 (3-nitrobenzeneboronic acid).

the pK of an ionizable group from the mandolonitrile pH profiles as there is no leveling off of rates at higher pH.

The profiles do give some indication that an ionizable group with higher pK may be involved. This may be the hydroxyl group of mandelonitrile, which has a pK of 10.73[33] The pK values of the boronic acids used range from 7 to 9. Boronic acid pK values are not evident in the pH profiles. It is also difficult to assign the pK of any group to the pH profile, since the pKs of mandelonitrile and boronic acids are not known at 60°C. It is known that the pK is temperature dependent, decreases with increasing temperature, and that the change in pK with temperature is linear.[34,35] But this change varies from one ionizable group to the other, which makes it difficult to know the exact decrease of the pK of mandelonitrile and boronic acid.

A straight line with a slope of − 0.4 is obtained when the logarithms of first-order rate constants for the hydrolysis by benzeneboronic acid are plotted as a function of pH. This suggests that the hydrolysis is not completely dependent on hydroxide ion, since the slope would have then been one. In the case of 3-nitrobenzeneboronic acid, the slope obtained is only − 0.17 (Figure 4). This suggests that the benzeneboronic acid-catalyzed reaction is more dependent on hydroxide ion than is the 3-nitrobenzeneboronic acid-catalyzed reaction.

Second-order rate constants are higher for benzeneboronic acids with electron-donating substituents than with electron-withdrawing substituents. On a

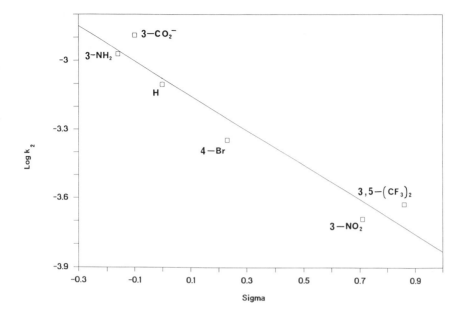

FIGURE 5. Hammett plot of the substituent effect on the hydrolysis of mandelonitrile (0.076 M) by benzeneboronic acids in 0.1 M bicarbonate buffer, pH 9.2, at 60°C. The ρ is −0.75. The ordinate has units M^{-1} sec^{-1}.

Hammett plot using the catalytic second-order rate constant, all boronic acids fall on a straight line with a slope of −0.75 (Figure 5). This value may result from binding components and from reaction rates that are contained in the overall second-order rate constants.

It is unlikely that boronic acid binding to mandelonitrile is improved by electron-donating substituents. The reverse is true for the complexation of boronates to sugars,[36] for the binding term seen in boronic acid-catalyzed hydrolysis of imines[23] and for the binding of boronic acids to serine proteases.[11,37] Thus, the negative ρ-value is likely to be due to the effect of substituents on a catalytic rate constant that may follow boronic acid binding to mandelonitrile.

Based on the above results, two possible mechanisms are proposed for the hydrolysis of mandelonitrile by the boronic acid.

A. PROPOSED MECHANISM (SCHEME 1)

In the first step, the boronic acid esterifies to mandelonitrile through the hydroxyl group of mandelonitrile and forms complex **2**, in which the boron atom is anionic and tetrahedral. Complex **2** is in equilibrium with complex **3**, which contains a trigonal boron atom. Boronic acids with electron-withdrawing substituents stabilize complex **2**. In the case of boronic acids with electron-donating substituents, complex **3** may be prevalent. Electron-donating groups tend to stabilize complex **3**. The hydrolytic mechanism is different

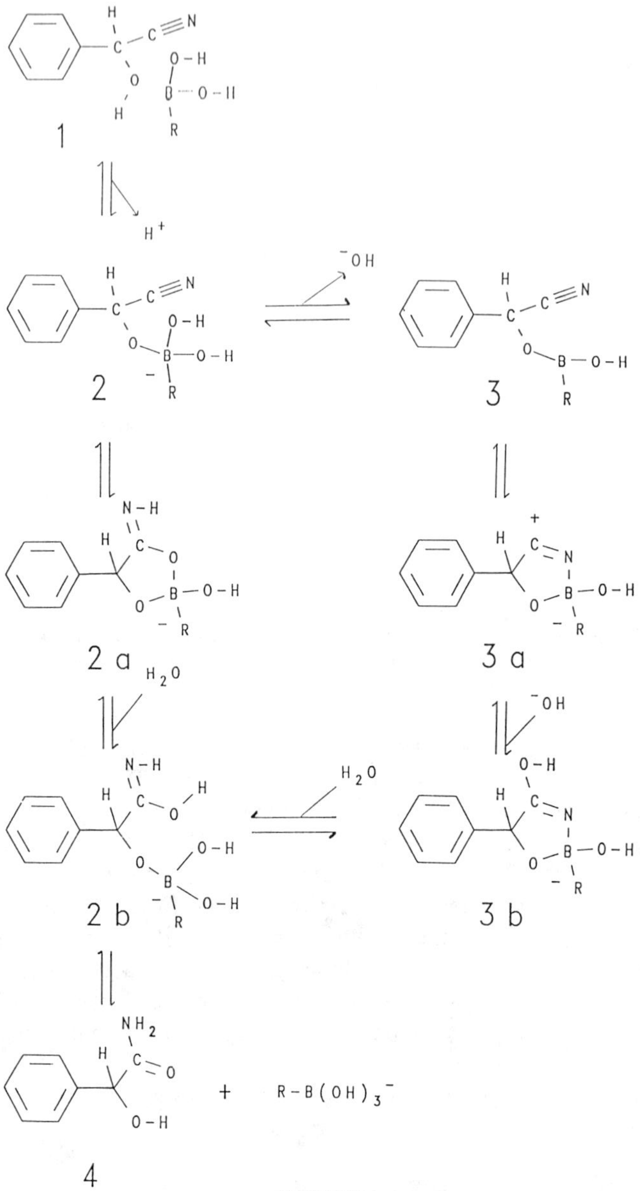

SCHEME 1.

for reactions that begin with these different complexes. Two competing mechanisms may best explain the different and noninteger relationships between pH and the logarithm of catalytic rate constants.

Dissociation constants for complexes **2** and **3** are likely to be very high since mandelonitrile is a monodentate ligand. Previous efforts to determine boronate dissociation constants for such ligands in aqueous media have failed.[32]

1. Path A (Leading from 2 to 3, 3a, 3b, 2b, and 4 in Scheme 1)

The boron atom in complex 3 acts as a Lewis acid and forms a five-membered ring, 3a. The boron-nitrogen bond makes the nitrile carbon atom more electron deficient. In the next step, the hydroxide ion attacks the carbonium ion. In subsequent steps, 3b undergoes hydrolysis followed by tautomerization to give products.

Metal ions hydrolyze phenanthroline nitrile to the corresponding amide using a mechanism similar to that proposed here.[38]

2. Path B (Leading from 2 to 2a, 2b, and 4 in Scheme 1)

In complex 2, a boronic acid oxygen atom nucleophilically attacks the nitrile carbon atom and forms a five-membered ring, 2a. In the next step, the ring undergoes hydrolysis and gives the intermediate 2b, which is same as the one formed in Path A. It is also possible that 2b forms through intermediate 3b. Intermediate 2a forms by the intramolecular transfer of a hydroxyl group from the boronic group to the nitrile carbon atom. Unlike Path A, Path B does not require hydroxide ion. In the case of benzeneboronic acid, complex 3 is likely to be the major species and follows Path A for the hydrolysis. The benzeneboronic acid-catalyzed hydrolysis is more dependent on hydroxide ion than that of 3-nitrobenzeneboronic acid, as it follows Path A for the hydrolysis. It is possible that the benzenboronic acid-catalyzed hydrolyses also follow Path B to some extent, as this hydrolysis is not completely dependent on hydroxide ion.

Complex 2 is likely to be the major species with 3-nitrobenzeneboronic acid, as it stabilizes the negative charge on boron which leads to Path B for the hydrolysis. 3-Nitrobenzeneboronic acid-catalyzed hydrolysis is not more dependent on hydroxide ion, as Path B does not require hydroxide ion for the hydrolysis. As the pH profile suggests, it might follow Path A for the hydrolysis to the smallest extent.

The Hammett plot (Figure 5) shows that benzeneboronic acids with electron-donating substituents hydrolyze mandelonitrile more rapidly. Benzeneboronic acids with electron-withdrawing substituents stabilize complex 2 compared to complex 3 and probably follow Path B for the hydrolysis. On the other hand, benzeneboronic acids with electron-donating substituents destabilize complex 3 compared to complex 2 and probably follow Path A. Reactions following Path A may be faster than those following Path B, resulting in higher rates for those benzeneboronic acids with electron-donating substituents.

REFERENCES

1. Edwards, J. O., Morrison, G. C., Ross, V., and Schultz, J. W., *J. Am. Chem. Soc.*, 77, 266, 1955.
2. Boeseken, J., *Adv. Carbohydr. Chem.*, 4, 189, 1949.
3. Steinberg, H. and Hunter, D. L., *Ind. Eng. Chem.*, 49, 174, 1957.
4. Kustin, K. and Pizer, R., *J. Am. Chem. Soc.*, 91, 317, 1969.
5. Friedman, S., Pace, B., and Pizer, R., *J. Am. Chem. Soc.*, 96, 5381, 1974.
6. Lorber, G. and Pizer, R., *Inorg. Chem.*, 15, 978, 1976.
7. Friedman, S. and Pizer, R., *J. Am. Chem. Soc.*, 97, 6059, 1975.
8. Pizer, R. and Babcock, L., *Inorg. Chem.*, 16, 1677, 1977.
9. Babcock, L. and Pizer, R., *Inorg. Chem.*, 19, 56, 1980.
10. Weith, H. L., Weibers, J. L., and Gilham, P. T., *Biochemistry*, 9, 4396, 1970.
11. Torssel, K., *Ark. Kemi*, 10, 529, 1957.
12. Philipp, M. and Bender, M. L., *Proc. Natl. Acad. Sci. U.S.A.*, 68, 478, 1971.
13. Kettner, C. A. and Shenvi, A. B., *J. Biol. Chem.*, 259, 15106, 1984.
14. Philipp, M., Claeson, G., Matteson, D. S., DeSoyza, T., Agner, E., and Sadhu, M., *Fed. Proc.*, 46, 2223, 1987.
15. Koehler, K. A. and Lienhard, G. E., *Biochemistry*, 10, 2477, 1971.
16. Peer, H. G., *Rec. Trav. Chim.*, 79, 825, 1960.
17. Jencks, W. P., *Catalysis in Chemistry and Enzymology*, McGraw-Hill, New York, 1969, 30.
18. Letsinger, R. L., Dandegaonker, S., Vullo, W. J., and Morrison, J. D., *J. Am. Chem. Soc.*, 85, 2223, 1963.
19. Letsinger, R. L. and Morrison, J. D., *J. Am. Chem. Soc.*, 85, 2227, 1963.
20. Letsinger, R. L. and MacLean, D. B., *J. Am. Chem. Soc.*, 85, 2230, 1963.
21. Capon, B. and Ghosh, B. C., *J. Chem. Soc. B*, 472, 1966.
22. Okuyama, T., Nagamatsu, H., and Fueno, T., *J. Org. Chem.*, 46, 1336, 1981.
23. Rao, G. and Philipp, M., *J. Org. Chem.*, 56, 1505, 1991.
24. Matsuda, H., Nagamatsu, H., Okuyama, T., and Fueno, T., *Bull. Chem. Soc. Jpn.*, 57, 500, 1984.
25. Hoffmann, J. and Sterba, V., *Collect. Czech. Chem. Commun.*, 37, 2043, 1972.
26. Nagamatsu, H., Okuyama, T., and Fueno, T., *Bull. Chem. Soc. Jpn.*, 57, 2502, 1984.
27. Nagamatsu, H., Okuyama, T., and Fueno, T., *Bull. Chem. Soc. Jpn.*, 57, 2508, 1984.
28. Wechsberg, M. and Schönbeck, R., *Alpha-Hydroxycarboxylic Acid Amides*, Austrian Patent 358,552, 1978; *Chem. Abstr.*, 94, 120904k, 1981.
29. Long, C., *Biochemists' Handbook*, Van Nostrand Reinhold, New York, 1961, 30.
30. Bean, F. R. and Johnson, J. R., *J. Am. Chem. Soc.*, 54, 4415, 1962.
31. Jorns, M. S., *Biochim. Biophys. Acta*, 613, 203, 1980.
32. Koehler, K. A., Jackson, R. C., and Lienhard, G. E., *J. Org. Chem.*, 37, 2232, 1972.
33. Ching, W. M. and Kallen, R. G., *J. Am. Chem. Soc.*, 100, 6119, 1978.
34. Chremos, G. N. and Zimmerman, H. K., *Chim. Chronika*, 28, 103, 1963.
35. Everett, D. H. and Pinsent, B. R. W., *J. Chem. Soc.*, 1029, 1948.
36. Torssell, K., McClendon, J. H., and Somers, G. M., *Acta Chem. Scand.*, 12, 1373, 1958.
37. Tsai, I. and Bender, M. L., *Arch. Biochem. Biophys.*, 228, 555, 1984.
38. Breslow, R., Fairweather, R., and Keana, J., *J. Am. Chem. Soc.*, 89, 2135, 1967.

Chapter 11

ACTIVATORS AND INHIBITORS OF CARBOXYPEPTIDASE A-CATALYZED HYDROLYSES

John F. Sebastian and Jeffrey L. Frye

TABLE OF CONTENTS

I. INTRODUCTION

Bovine carboxypeptidase A (CPA) is a zinc-containing exopeptidase that catalyzes the hydrolysis of polypeptides at the C-terminal residue.[1-8] The enzyme also catalyzes hydrolysis of oligopeptides, N-acyldipeptides, thioamides[9] and thioesters,[9] α-β-elimination reactions,[10] and stereospecific exchange of one of the methylene α-hydrogens of a ketone.[11,12] The α-carbon of the C-terminal residue of peptide substrates must have the L-configuration, and there is a preference for bulky hydrophobic side chains. The active site of CPA contains a dead-end pocket lined by apolar side-chain residues. In addition to the pocket, groups implicated in binding and catalysis include Asn-144, Arg-145, Zn^{2+}, Glu-270, Tyr-248, Arg-127, and Arg-71. Five substrate binding subsites (S_1', S_1 through S_4), each accommodating one amino acid residue of the substrate, have been reported.[13] A secondary binding locus for modifiers and short peptide or ester substrates, involving CPA residues Tyr-198, Phe-279, and Arg-71, has been suggested.[14]

The mechanism of action of CPA catalysis of peptide hydrolysis is the subject of much controversy. Recently, a mechanism has been proposed which favors a promoted water pathway.[15] In this mechanism, nucleophilic attack by a water molecule promoted by zinc and assisted by Glu-270 (as a proton acceptor) results in formation of a tetrahedral intermediate stabilized by a variety of groups contributed by the enzyme. Arg-127 achieves prominence by polarizing the carbonyl oxygen of the scissle peptide bond, stabilizing the oxyanion of the tetrahedral intermediate, and hydrogen bonding to a zinc-bound carboxylate in the enzyme-products complex.

Another subject of considerable debate involves possible differences in the mechanisms of esterase and peptidase activities, and in the binding sites for peptide and ester substrates. There is evidence indicating that at least two intermediates are involved in the kinetic mechanisms of both types of substrate, but that the rate-determining steps of the reactions are different.[16-19] Furthermore, it has been suggested[15] that differences between proteolysis and esterolysis are due to different roles for zinc and Arg-127; presumably, zinc rather than Arg-127 polarizes the carbonyl of the ester bond, although recent site-directed mutagenic studies show that Arg-127 is required for peptide *and* ester hydrolysis.[20] Others have suggested that the substrate binding sites for peptide and ester are quite different.[21]

We have taken a rather different approach to the problem. We have chosen to study the N-acyldipeptide substrate N-benzoylglycyl-L-phenylalanine (BGP) and the corresponding ester substrate N-benzoylglycyl-L-phenylacetate (BGPL) because (1) so much work has been done with these substrates, (2) they are sufficiently short so that when bound to the catalytic site, the putative modifier binding site — formed by Tyr-198, Phe-279, Arg-71, and the bound substrate — will remain exposed, and (3) from the carbon alpha to the scissile peptide or ester bond to the N-benzoyl group, the two substrates are identical. Thus, any differences in the way esters and peptides bind to the enzyme might be reflected by the effects of modifiers binding to the ES complex. These differences would presumably arise (and perhaps might be amplified) by the different binding and catalytic requirements of the reactive peptide and ester moieties. Many kinetic studies have focused on the effects of reversible inhibitors exhibiting primarily competitive or mixed inhibition. We decided to study another class of modifiers: those that accelerate peptide hydrolysis but uncompetitively inhibit ester hydrolysis.

Finally, we wish to turn to a potentially interesting aspect of modifier structure. Adamantane carboxylate is known to form a strong inclusion complex with β-cyclodextrin.[22] Molecular models suggest that the hydrophobic pocket of CPA seems to be of the appropriate size and shape to accommodate the spherically shaped adamantyl group. It occurred to us that the hydrophobic interaction between the adamantyl moiety and the pocket might contribute to stronger binding of an inhibitor to the enzyme. Possible candidates for study might include adamantane acetate, 3-adamantane propanoate, 1,3-adamantane diacetate, and the diacid salt obtained by substitution of phenyl by adamantyl in 2-benzylsuccinate. We report here some preliminary results obtained with adamantane acetate.

II. EXPERIMENTAL

A. CPA

CPA_α (Cox) was purchased from Sigma Chemical Co. The stated activity of the toluene suspension of CPA crystals was 57 U/mg, with a concentration of 20 mg/ml. The toluene suspension was mixed using a vortex mixer and an aliquot was removed and centrifuged at 10,000 rpm for 15 min. The liquid was removed and the crystals were washed twice with double-distilled water and dissolved in 3.0 M NaCl. After dilution to the desired volume, this stock solution was stored at 4°C.

B. BUFFER

Reactions were carried out in a buffer solution of 1.0 M NaCl-0.05 M Tris. The pH was measured using a Beckman Expandomatic pH meter and adjusted to 7.50 with 1.0 M NaOH or HCl. Solutions were stored at 4°C.

C. KINETIC MEASUREMENTS

Kinetic data were obtained on a Cary-14 recording spectrophotometer equipped with a thermostatted sample chamber. The reaction conditions for all samples were pH 7.50, which was checked immediately before each new solution was used, and 25.0 ± 0.1°C, which was maintained with a Lauda circulator. The reaction was followed at 254 or 258 nm.

In a typical run, 3.00 ml of the substrate solution was transferred via pipet into a 1.00-cm quartz cell equipped with a stopper. This cell was then placed in the thermostatted cell compartment of the Cary-14 spectrophotometer and allowed to equilibrate at 25.0°C for 10 to 15 min. An identical sample was placed in the reference chamber and used as a blank. To start each sample, 25 or 50 μl of the standard enzyme solution was delivered by means of a calibrated micropiepet to a flat-tipped stirring rod, which was then used to transfer the enzyme to the cuvette at zero time. The reaction mixture was stirred rapidly for 5 to 10 s, at which point the spectrophotometer was activated and the elapsed time was recorded. After monitoring the change in absorbance for 2 to 4 min, another sample was started in the same manner. Each final absorbance (A_∞) was generally read within 1 h after the sample was started.

A first-order kinetic plot was constructed from absorbance and time data taken directly from the spectrophotometer chart paper by using the following equations.

$$\ln \frac{C_t}{C_0} = -kt$$

where C_0 and C_t represent the concentration of substrate at time = 0 and t, respectively. Since $A = \epsilon Cl$, where l is the path length and ϵ is the molar absorptivity, and both l and ϵ are constant, this can be rewritten in terms of absorbance units:

$$\ln \frac{A_\infty - A_t}{A_\infty - A_0} = -kt$$

$$\ln(A_\infty - A_t) = -kt + \ln(A_\infty - A_0)$$

where k is the first-order rate constant. A_∞, A_t, and A represent the infinite absorbance, the absorbance at time t, and the absorbance at time 0, respectively. ($A_\infty - A_0$) and $A_\infty - A_t$) represent the initial substrate concentration and the substrate concentration at time t, respectively. A least-squares computer program was used to calculate the rate constants and initial velocities.

All of the experimental initial velocity data calculated from the least-squares computer program were plotted in double-reciprocal form to check on linearity. The kinetic parameters K_m and V_{max} were obtained from a nonlinear least-squares fit of the Michaelis-Menten equation.

D. SUBSTRATES

BGP was purchased from Sigma Chemical Co., stored at or below 0°C, and used without further purification. A typical stock solution of $2.0 \times 10^{-3} M$ was prepared in 0.05 M Tris-1.0 M NaCl buffer and standardized to pH 7.50 with NaOH or HCl. When not in use, all solutions containing substrate were stored at 4°C.

BG-L-PL was purchased from Bachem Inc., and BG-DL-PL was purchased from Sigma Chemical Co. Both were stored at 0°C and used without further purification. Irrespective of source of substrate used in these investigations,

TABLE 1

Effect of Modifiers on the CPA-Catalyzed Hydrolysis of BGP and BGPL

Modifier	Effect of modifier with		K_A (M)	K_I (M)	Ref.
	BGP	BGPL			
3,3-Diphenylpropanoate	A[a]	UCI[b]	2.1×10^{-3}	2.1×10^{-3}	28
N-Benzoylglycine	A[c]	UCI	—	8.1×10^{-3}	24, 25
N-Carbobenzyloxyglycine	A[c]	UCI	1.7×10^{-2} [d]	8.2×10^{-3}	24—26
2,2-Dimethyl-2-silapentane-sulfonate	A	UCI	5.0×10^{-3}	1.6×10^{-3}	27, 29
2-Phenylethanesulfonate	A	UCI	2.1×10^{-3}	3.7×10^{-3}	23, 29

[a] Activation.
[b] Uncompetitive inhibition.
[c] Substrate was carbobenzyloxyglycyl-L-phenylalanine.
[d] We have estimated this value from results given in Reference 26.

the range of concentrations of the L-enantiomer was 3.71×10^{-5} to 1.98×10^{-4} M. A stock solution of 3.00×10^{-4} M (L-enantiomer) was prepared in 0.05 M Tris-1.00 M NaCl buffer, pH 7.50.

E. MODIFIERS

Modifiers were either purchased or synthesized as described previously.[23]

III. RESULTS AND DISCUSSION

The effects of 3,3-diphenylpropanoate (DPP), 2,2-dimethyl-2-silapentane-sulfonate (DSS), 2-phenylethanesulfonate (PES), benzoylglycine (BG), and carbobenzyloxglycine (CbzG) on the CPA-catalyzed hydrolysis of short peptide and ester substrates are shown in Table 1. The results for BG,[24,25] CbzG,[24-26] and of activation by DSS[27] were determined by others. In every case for which detailed kinetics are available, activators of peptide hydrolysis are uncompetitive inhibitors of ester hydrolysis.

The simplest interpretation of these results is that all of the modifiers listed bind to the secondary pocket (or cleft) and that, due to somewhat different binding sites for at least that portion of the substrate extending outward from the sissile bond region to the N-acyl blocking group, the effect of a modifier is different with each of the two types of substrates. The details of the origin of these effects are unclear. However, if it is assumed that the carbonyl group of the sissile bond is oriented somewhat differently for peptide substrates than for the esters, as has been suggested,[15] it seems reasonable to assume that the N-benzoylglycyl groups of the two substrates would not necessarily bind in precisely the same site.[30] Furthermore, given the rather low barrier to rotation about the ester C-O bond and the sensitivity of the rate of ester hydrolysis to ester conformation,[31,32] inhibition might result if the bound

TABLE 2
The Effect of Adamantane Acetate on
the CPA-Catalyzed Hydrolysis of BGPL

Mode	K_I (M) (for IE)	K_{II} (M) (for IES)
Mixed	$(8 \pm 2) \times 10^{-3}$	$(2.8 \pm 0.2) \times 10^{-3}$

modifier prevents or in some way interferes with the achievement of a catalytically active conformation by the ester C-O bond. Acceleration of the hydrolysis of the more rigid peptide bond could be achieved by increasing the concentration of catalytically active complexes (ES and AES) due to the binding of the activator to the ES complex, even if ES and AES collapse to products with the same rate constant.

Adamantane acetate was found to be a mixed inhibitor of ester hydrolysis. The results are given in Table 2. The comparatively large values of the inhibition constants and the observation that the modifier is a mixed rather than a competitive inhibitor indicates that no special advantage results from the presence of the adamantyl group in this acetate derivative. This result can be compared to that of cyclohexylacetate which has been reported to give partial competitive inhibition of BGPL hydrolysis with a K_I of 1.2×10^{-2} M for the EI complex and K_{II} of 1.0×10^{-3} M for the EI$_2$ complex.[33]

IV. CONCLUSION

We wish to point out that many of the techniques and principles employed in this work were pioneered by Myron Bender. His application of physical organic chemistry to enzymology has given us considerable insight into how enzymes work.

ACKNOWLEDGMENT

We wish to thank Dr. Charles Griffin for several computer programs.

REFERENCES

1. **Quiocho, F. A. and Lipscomb, W. N.,** *Adv. Protein Chem.,* 25, 1, 1971.
2. **Hartsuck, J. A. and Lipscomb, W. N.,** in *The Enzymes,* Vol. 3, Boyer, P., Ed., Academic Press, New York, 1971, 1.
3. **Bradshaw, R. A., Ericsson, L. H., Walsh, K. A., and Neurath, H.,** *Proc. Natl. Acad. Sci. U.S.A.,* 63, 1389, 1969.
4. **Lipscomb, W. N.,** *Proc. Natl. Acad. Sci. U.S.A.,* 77, 3875, 1980.
5. **Lipscomb, W. N.,** *Acc. Chem. Res.,* 15, 232, 1982.
6. **Lipscomb, W. N.,** *Annu. Rev. Biochem.,* 52, 17, 1983.

7. **Vallee, B. L., Gales, A., Auld, D. S., and Riordan, J. F.**, in *Metal Ions in Biology*, Vol. 5, Spiro, T. G., Ed., John Wiley & Sons, New York, 1983, 25.
8. **Vallee, B. L. and Galdes, A.**, *Adv. Enzymol.*, 56, 283, 1984.
9. **Campbell, P. and Nashed, N. T.**, *J. Am. Chem. Soc.*, 104, 5221, 1982.
10. **Nashed, N. T. and Kaiser, E. T.**, *J. Am. Chem. Soc.*, 103, 3611, 1981.
11. **Sugimoto, T. and Kaiser, E. T.**, *J. Am. Chem. Soc.*, 100, 7750, 1978.
12. **Sugimoto, T. and Kaiser, E. T.**, *J. Am. Chem. Soc.*, 101, 3946, 1979.
13. **Abramowitz, N., Schechter, I., and Berger, A.**, *Biochem. Biophys. Res. Commun.*, 29, 862, 1967.
14. **Bunting, J. W. and Kabir, S. H.**, *J. Am. Chem. Soc.*, 99, 2775, 1977.
15. **Christianson, D. W. and Lipscomb, W. N.**, *Acc. Chem. Res.*, 22, 62, 1989.
16. **Auld, D. S., Galdes, A., Geoghegan, K. F., Holmquist, B., Martinelli, R. A., and Vallee, B. L.**, *Proc. Natl. Acad. Sci. U.S.A.*, 81, 5041, 1984.
17. **Geoghegan, K. F., Galdes, A., Marinelli, R. A., Holmquist, B., Auld, D. S., and Vallee, B. L.**, *Biochemistry*, 22, 2255, 1983.
18. **Georghegan, K. F., Galdes, A., Hanson, G., Holmquist, B., Auld, D. S., and Vallee, B. L.**, *Biochemistry*, 25, 4669, 1986.
19. **Galdes, A., Auld, D. S., and Vallee, B. L.**, *Biochemistry*, 25, 646, 1986.
20. **Phillips, M. A., Fletterick, R., and Rutter, W. J.**, *J. Biol. Chem.*, 265, 20692, 1990.
21. **Vallee, B. L. and Galdes, A.**, *Adv. Enzymol. Relat. Areas Mol. Biol.*, 56, 369, 1984.
22. **VanEtten, R. L., Sebastian, J. F., Clowes, G. A., and Bender, M. L.**, *J. Am. Chem. Soc.*, 89, 3242, 1967.
23. **Sebastian, J. F., Hinks, R. S., and Reuland, R. V.**, *Biochem. Cell Biol.*, 65, 717, 1987.
24. **Davies, R. C., Riodan, J. F., Auld, D. F., and Vallee, B. L.**, *Biochemistry*, 7, 1090, 1968.
25. **Bunting, J. W. and Myers, C. D.**, *Can. J. Chem.*, 53, 1993, 1975.
26. **Whitaker, J. R.**, *Biochem. Biophys. Res. Commun.*, 22, 6, 1966.
27. **Epstein, M. and Navon, G.**, *Biochem. Biophys. Res. Commun.*, 36, 126, 1969.
28. **Sebastian, J. F. and Lo, W.-Y.**, *Can. J. Biochem.*, 56, 329, 1978.
29. **Frye, J. L. and Sebastian, J. F.**, *Biochem. Cell Biol.*, 68, 1062, 1990.
30. **Frye, J. L. and Sebastian, J. F.**, unpublished results.
31. **Menger, F. M. and Sherrod, M. J.**, *J. Am. Chem. Soc.*, 110, 8606, 1988.
32. **Thiem, H.-J., Brandl, M., and Breslow, R.**, *J. Am. Chem. Soc.*, 110, 8612, 1988.
33. **Bunting, J. W. and Myers, C. D.**, *Biochim. Biophys. Acta*, 341, 222, 1974.

Chapter 12

MEDIATION OF ENZYME CATALYSIS BY INTRA- AND INTERSITE CHANGES IN CONFORMATIONAL STATE

Frederick C. Wedler

TABLE OF CONTENTS

I. INTRODUCTION

While I was a graduate student in the Bender laboratory, 1964 to 1968, the classical "lock-and-key" model for substrate binding to enzyme active sites was rapidly losing ground to Koshland's "induced fit" model. Systems such as aspartate transcarbamylase provided dramatic evidence that protein structures were dynamically plastic, and could also exist in different discrete conformational states. This also applied to monomeric systems such as the serine proteases, e.g., the decrease in substrate affinity, $1/K_m$ (app), for chymotrypsin at alkaline pH caused by disruption of an Asp-194-Ile-16 hydrogen bond, essential for the structural integrity of the active site.[1,2] This raised several questions. Could catalysis (k_{cat}) as well as binding be altered by conformational changes that were dependent on pH and temperature? What were the minimal steric requirements of substrate to drive reconversion to the high-affinity state? Finally (to test the dogma that "better binding = better catalysis"), to what extent were binding and catalysis connected?

Meanwhile, during 2 postdoctoral years at UCLA, Paul Boyer introduced me to glutamine synthetase and equilibrium isotope exchange kinetics. In this environment, I became aware that, for a variety of intriguing multisubstrate enzymes, the substrate binding curves were nonhyperbolic and that site-site interactions were clearly crucial in multimeric enzymes that play key roles in metabolic regulation.

Over the past 2 decades, we have designed experiments to probe the mechanistic basis for substrate-induced conformational changes in enzymes, which *a priori* can cause changes in the rate-limiting step, kinetic mechanism, substrate affinity, and the rate of net turnover. Further, catalysis may be enhanced or optimized by conformational changes induced by substrate binding to multimeric systems ("catalytic cooperativity"), with transfer of the energy derived from substrate binding from one site to another in order to facilitate catalysis or the rate of product release.

II. CHYMOTRYPSIN

My first assignment in the Bender lab was to study the "aging reaction" of phosphate ester-inhibited serine proteases.[3,4] In this Vietnam era, however, the potential application of this reaction to nerve gas development made me pleased to move to another problem. The affinity of chymotrypsin for substrates had been shown to decrease sharply at alkaline pH upon ionization of a single group, the α-NH_4^+ group of Ile-16 at the N terminus of the B-chain,[1] shown in the crystal structure to form a salt bridge to Asp-194, thus helping to align Ser-195 properly for catalysis.[2] In the minimal model, substrate binds preferentially to protonated enzyme, so that substrate binding at or above the pK_a will result in proton uptake:

$$E \underset{K_a}{\overset{H^+}{\rightleftharpoons}} EH^+ \underset{K_s}{\overset{S}{\rightleftharpoons}} EH^+S$$

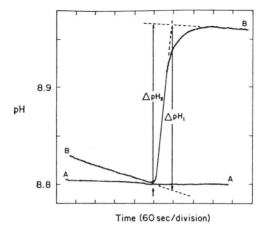

FIGURE 1. A typical trace for the recording pH meter apparatus, observed pH against time, involving 10^{-4} M chymotrypsinogen A (curve A) and 10^{-4} M α-chymotrypsinogen A (curve B), 0.1 M KCl, upon addition of benzyl alcohol (K_i = 10 mM, pH 7.9), to give a final concentration of 200 mM; 25°C. The "observed ΔpH" was calculated using the graphically determined values and $ΔpH_{obs} = (ΔpH_1 + ΔpH_2)/2$, from which could be calculated a value for n (see text). (Reprinted with permission from Wedler, F. C. and Bender, M. L., *J. Am. Chem. Soc.*, 94, 2101, 1972. Copyright 1972, American Chemical Society.)

This hypothesis was proven by the experimental observation of proton absorption (Figure 1) with pK_a (app) = 8.8, as shown in Figure 2.[4]

How large a side-chain group is required for K_m (app) to be altered by the conformational change at alkaline pH, or, conversely, what is the minimum size required to reverse this conformational change and induce the correct conformation of Ser-195 required for catalysis (k_{cat})? To address these questions, the effects of pH on k_{cat} and K_m (app) were determined for Leu- and Ala-ester substrates with both α- and δ-forms of chymotrypsin.[6] As shown in Figure 3, $1/K_m$ decreases above pH 8 for both Leu and Ala substrates, but k_{cat} does not. Since k_{cat} is 100 times lower for Ala than for Leu, apparently the isobutyl group of Leu induces an active-site conformation more favorable for catalysis than that afforded by the methyl group of Ala.

Since temperature can also shift the equilibria for both pK_as and protein conformations, it was important to determine the effects of temperature on k_{cat} and K_m. Despite extensive kinetic data for chymotrypsin, never before had such studies been performed on a systematic basis. Amino acid methylester substrates with side-chain groups of different sizes (Phe, Leu, and Ala) were studied at pH values above and below pK_a 8.8 for the alkaline conformational transition in chymotrypsin binding, as shown in Figure 4.[7]

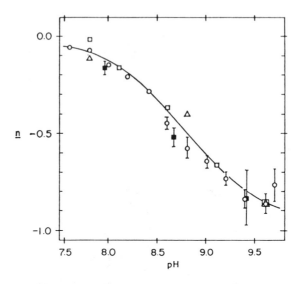

FIGURE 2. The pH dependence of proton absorption, n, by 10^{-4} M α-chymotrypsin upon binding of inhibitors. pH-recorder data: benzyl alcohol (○), acetonitrile (△), N-acetyl-D-tryptophan-amide (□), $[I_0]$ always five times > K_i at each pH; 25°C. pH-indicator dye technique with thymol blue and phenolphthalein: α-chymotrypsin, 10^{-4} M, and benzyl alcohol, ■. The solid line is calculated from Equation 5, assuming pK_a (app) = 8.8. (Reprinted with permission from Wedler, F. C. and Bender, M. L., *J. Am. Chem. Soc.*, 94, 2101, 1972. Copyright 1972, American Chemical Society.)

Activation and thermodynamic parameters calculated from these plots provide important clues to the underlying mechanism or basis for these effects, namely, that entropic factors are quite important. Consistent with this, it was observed that active-site specificity *decreases* with an increase in temperature or pH, a decrease in the size of the amino acid substrate side-chain group, or upon conversion from δ- to α-chymotrypsin — all of which increase disorder or conformational degrees of freedom in the active site of the enzyme-substrate complex or the acyl-enzyme intermediate.

For the overall mechanism of chymotrypsin, involving formation of an acyl-enzyme intermediate (E-P_2),

$$E + S \underset{k_{-1}}{\overset{k_1}{\rightleftharpoons}} E{\cdot}S \underset{k_{-2}}{\overset{k_2}{\rightleftharpoons}} \underset{+P_2}{E{-}P_2} \underset{k_{-3}}{\overset{k_3}{\rightleftharpoons}} E + P_2$$

Since $k_2 > k_3$, the rate-limiting step with ester substrates at 25°C and pH 7.8 is deacylation (k_{+3}). Changes in slope observed in Figure 4 with Arrhenius plots (catalysis) were much less frequent and less dramatic than those observed

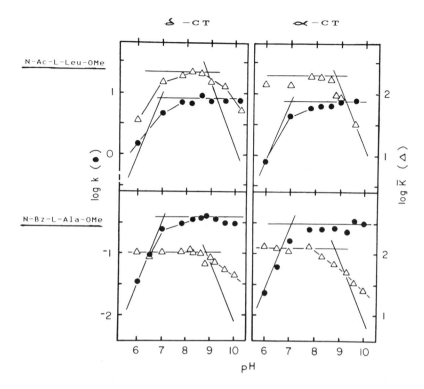

FIGURE 3. pH dependence of k_{cat} and $1/K_m$ (K) for δ- and α-chymotrypsin with methyl-ester substrates of Ala and Leu. (Reprinted from Wedler, F. C., Uretsky, L. S., and McClune, G., *Biochim. Biophys. Acta,* 370, 541, 1974. With permission.)

for the van't Hoff plots (binding). A temperature-dependent change in the rate-determining step from k_3 to k_2 should cause nearly parallel changes in both types of plots. This suggests that pH and temperature primarily alter the binding (k_1, k_{-1}) step, i.e., the conformation of the substrate binding site which changes as a two-state process, dependent upon ionization of the α-NH_4^+ group of Ile-16. Once the ES complex has formed, the degree to which Ser-195 is optimally aligned for formation and breakdown of acyl enzyme (i.e., the turnover rate) is further finetuned by the steric features of the substrate side-chain group in its subsite.

III. GLUTAMINE SYNTHETASE

A. *ESCHERICHIA COLI*

In *E. coli,* catalysis of L-Gln formation by glutamine synthetase:

$$\text{L-Glu} + NH_3 + \text{MgATP} \rightarrow \text{L-Gln} + \text{MgADP} + P_i$$

(1)   Phe

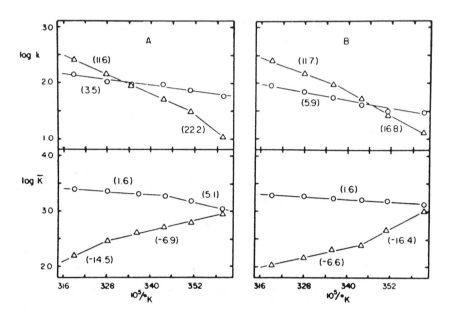

FIGURE 4.   Temperature dependence of catalysis and binding for (A) δ and (B) α at pH 7.8 (○) and 9.6 (△), with *N*-acyl methyl-ester substrates of (1) Phe, (2) Leu, and (3) Ala. (Reprinted from Wedler, F. C., Uretsky, L. S., McClune, G., and Cencula, J., *Arch. Biochim. Biophys.*, 170, 476, 1975. With permission.)

is stringently regulated by feedback inhibition with end-product metabolites derived from L-Gln.[8] Surprisingly, although this enzyme is composed of 12 subunits, little evidence of subunit interactions has been found. Then it was observed that the irreversible inhibition of GS upon phosphorylation of the transition-state analog, L-Met-*S*-sulfoximine (Met-SOX), in the presence of MgATP showed strong deviations from first-order behavior.[9] Further, Lineweaver-Burk plots of initial velocity kinetics for ammonia binding were biphasic, indicating negative cooperativity.[10,11] Equilibrium binding of Met-SOX was a strongly biphasic process, with a set of high-affinity sites extrapolating to 0.4 to 0.5 and low-affinity sites to 1.0 mol of Met-SOX bound per subunit.[11] Most importantly, the rate of the Met-SOX + MgATP reaction was found to be sigmoidal, with Hill n_H = 2.2 and $S_{0.5}$ = 25 μ*M*, compared to K_{is} ≈ 1 μ*M* for Met-SOX as a competitive inhibitor of GS vs. L-Glu.[11] The fact that the binding constant for Met-SOX was 200-fold tighter than that for Met-sulfone[11] suggested that the imino group of Met-SOX contributed in a critical way to the interaction of GS with the transition state. To test this, we prepared GS with different numbers of subunits per dodecamer inactivated with Met-SOX + MgATP, as shown in Scheme 1. The effects of partial

(2) Leu

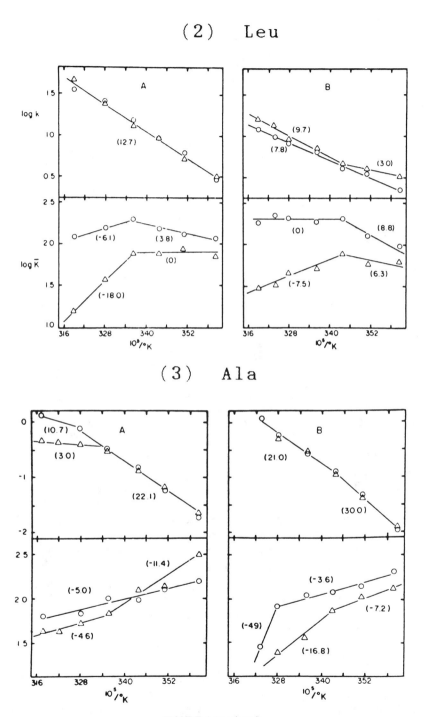

(3) Ala

FIGURE 4 (continued)

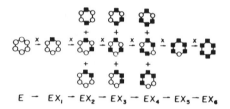

$$E \rightarrow EX_1 \rightarrow EX_2 \rightarrow EX_3 \rightarrow EX_4 \rightarrow EX_5 \rightarrow EX_6$$

SCHEME 1. (Reprinted with permission from Wedler, F. C., Sugiyama, Y., and Fisher, K. E., Biochemistry, 21, 2168, 1982. Copyright 1982, American Chemical Society.)

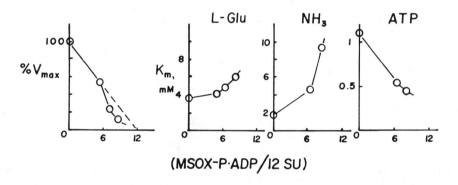

(MSOX-P·ADP/12 SU)

FIGURE 5.   Effects of partial inactivation on V_{max} and substrate K_m values with partially active forms of GS. (Reprinted with permission from Wedler, F. C., Sugiyama, Y., and Fisher, K. E., *Biochemistry,* 21, 2168, 1982. Copyright 1982, American Chemical Society.)

inactivation on V_{max} and substrate K_m values with these partially active forms of GS are shown in Figure 5.

V_{max} and K_m (ATP) both decrease linearly to 0.5 Met-SOX per subunit, then show deviations from a linear response. The K_m Glu increases only by about 50% in enzyme having only 15% activity, but that for NH_3 increases nonlinearly by fivefold, clearly the strongest effect observed. Taken together, these data led us to propose a model in which Met-SOX binding energy was used to drive phosphorylation by ATP at an adjacent subunit. The mechanism shown in Scheme 2 involves rotation of the sulfimino (S=NH) group out of the subsite ammonia in order for reaction with the γ-phosphoryl of ATP to occur. An equally plausible mechanism could be written in which the methyl or S=O groups of Met-SOX bind in the NH_3 subsite and the binding energy (Δ CONF.) step involves compression or movement of the imino group toward the γ-P of ATP.

The implications of these findings for the mechanism of GS are summarized in Scheme 3. Energy from the binding of NH_3 at one subunit is used to force the collapse of the highly stabilized transition state and to drive the release

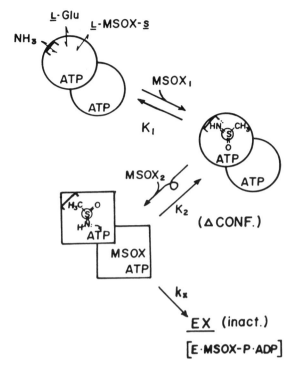

SCHEME 2. (Reprinted with permission from Wedler, F. C., Sugiyama, Y., and Fisher, K. E., Biochemistry, 21, 2168, 1982. Copyright 1982, American Chemical Society.)

of products. Previous data from equilibrium isotope-exchange kinetics studies[12,13] showed the latter step, not catalysis, to be the slow step in net turnover.

B. *BACILLUS CALDOLYTICUS*

The 6-8 modifier binding sites found per subunit for *E. coli* GS indicates evolution of a high density of functional information per 50,000-Da chain. This led us to wonder how much of this could be retained in GS adapted to life under extreme conditions, e.g., high temperatures. Isolation of GS from the highly thermophilic organism, *Bacillus caldolyticus,* led to the discovery that this organism synthesizes GS in two forms, called E-I and E-II, apparently as two separate gene products.[14] Based on the fact that E-I and E-II respond to different effectors in a complementary manner,[15] one reason for this appears to be a "division of labor" to allow total GS activity response to the full range of feedback modifiers with a reduction in the number of modifier sites per subunit, which would allow more protein structure to be devoted to thermostabilization. Another reason for this is based on the fact that E-I and E-II are differentially activated by Mg(II) and Mn(II) ions, respectively: for

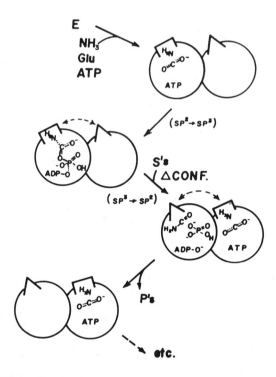

SCHEME 3.   (Reprinted with permission from Wedler, F. C., Sugiyama, Y., and Fisher, K. E., Biochemistry, 21, 2168, 1982. Copyright 1982, American Chemical Society.)

vegetative growth with nonlimiting nutrients, the relatively unregulated E-I activity may dominate, but with limiting nutrients upon initiation of the sporulation process, Mn(II) uptake occurs, which causes modifier-sensitive E-II activity to dominate.

What degree of complexity and plasticity can be maintained in a protein that is thermostable above 90°C?[16] The Arrhenius plots shown in Figure 6 indicate changes in E_a around 55 to 65°C for both E-I and E-II, with the most striking differences for E-II.

Other evidence for conformational flexibility of E-I and E-II was provided by studies of kinetic saturation curves for substrates at 70°C.[14] These data, shown in Figure 7, indicate that with the optimal activating cation present (Mg^{2+} with E-I, Mn^{2+} with E-II), biphasic Eadie-Hofstee plots indicative of negative cooperativity are observed for NH_3 but not the other substrates, except with L-Glu with Mn · E-I.

Based on our findings and hypotheses with *E. coli* GS outlined in the previous section, these experimental data suggest a mechanistic role for the effects in Figure 7, namely, "catalytic cooperativity"[17] may exist for thermophilic GS as well.

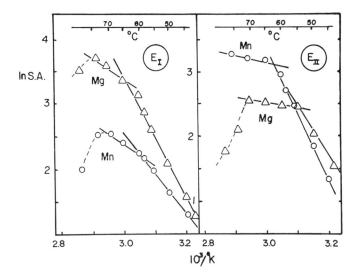

FIGURE 6.   Arrhenius plots for E-I and E-II forms of *B. caldolyticus* GS, activated by (○) Mn^{2+} or (△) Mg^{2+} ions. (Reprinted from Wedler, F. C., Shreve, D. S., Kenny, R. M., Ashour, A. E., Carfi, J., and Rhee, S. G., *J. Biol. Chem.*, 255, 9507, 1980. With permission.)

IV. ASPARTATE TRANSCARBAMYLASE

Gross changes in quaternary structure that accompany the allosteric transition in *E. coli* ATCase, which finetunes the binding of L-aspartate near 12 mM, have been studied intensely for 3 decades.[18] Catalysis of this first committed step in pyrimidine biosynthesis is regulated with feedback inhibition by end-product CTP and activation by ATP from the purine biosynthetic pathway:

$$C\text{-}P + \text{L-Asp} \rightleftharpoons N\text{-carbamyl-L-Asp} + P_i$$

Another intriguing question is how the energy of binding effectors to regulatory (r-) subunits is transmitted over 40 Å to the active sites, formed at the interface between catalytic (c-) chains.

In the course of using equilibrium isotope-exchange kinetics to verify that the kinetic mechanism for ATCase was compulsory order, it was reported that at 5°C the [^{14}C]-Asp $\rightleftharpoons$ C-Asp and [^{32}P]C-P $\rightleftharpoons$ P_i exchange rates were equal, which suggests that catalysis is rate limiting.[19] At 28°C, however, these two exchange rates were vastly different, and the saturation patterns were those expected for compulsory order binding of C-P before Asp, and C-Asp release before P_i.[20] This led us to determine the temperature dependence of the initial velocity with both holoenzyme (c_6r_6) and catalytic subunits (c_3), shown in Figure 8. Near 15°C, the E_a changes from 6.3 to 22 kcal/mol for

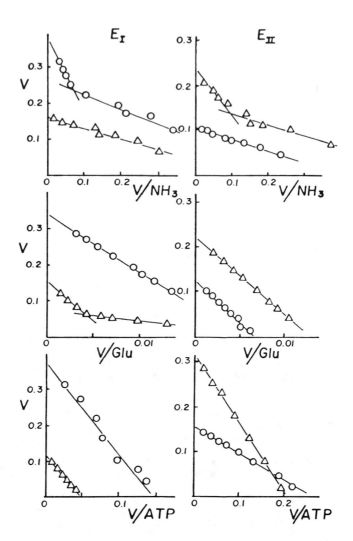

FIGURE 7.   Eadie-Hofstee plots of substrate binding with *B. caldolyticus* E-I and E-II forms of GS, activated by (○) Mn^{2+} or (△) Mg^{2+}. (Reprinted from Wedler, F. C., Shreve, D. S., Kenny, R. M., Ashour, A. E., Carfi, J., and Rhee, S., G., *J. Biol. Chem.*, 255, 9507, 1980. With permission.)

holoenzyme, but the plot for c_3 is strictly linear. Dissociation of c_6r_6 to c_3 causes the cooperative binding of Asp to become hyperbolic. This, along with the lack of a primary isotope effect on velocity with bridge-[^{18}O]-labeled C-P,[21] suggests that the T-R transition is the rate-limiting step in the forward direction. Comparison of initial velocity data to relative rates of isotope exchange reactions led to the conclusions in Scheme 4.

The mechanism of action of effectors on ATCase has been much studied, but only recently understood in some detail. Since the rate of the T-R transition

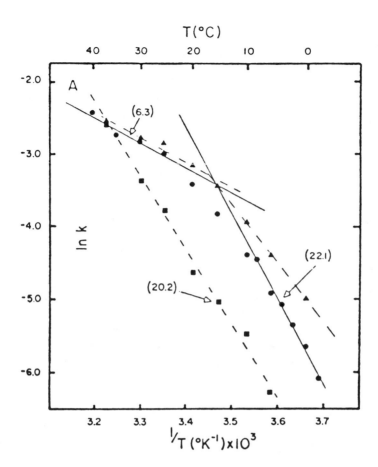

FIGURE 8. Temperature dependence of initial velocity of exchange rates with holoenzyme (c_6r_6 [▲,●]) and catalytic subunits (c_3 [■]). (Reprinted from Wedler, F. C. and Gasser, F. J., *Arch. Biochim. Biophys.*, 163, 57, 1974. With permission.)

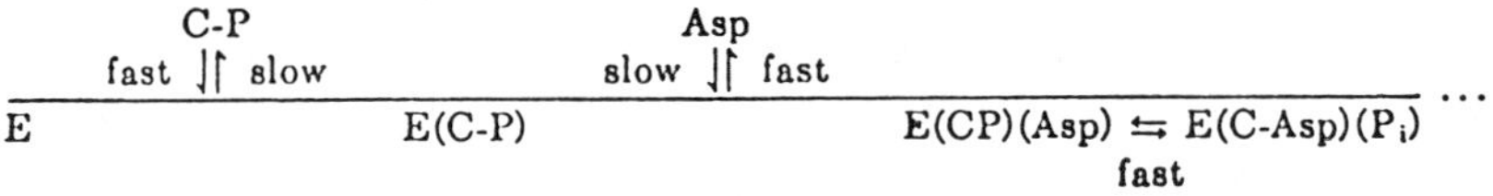

SCHEME 4. (Reprinted from Wedler, F. C. and Gasser, F. J., *Arch. Biochim. Biophys.*, 163, 57, 1974. With permission.)

apparently determines V_{max} for the forward reaction, and since the effectors (CTP and ATP) alter $S_{0.5}$ (Asp), but not V_{max} to any appreciable degree, one can argue that effectors do not change the rate of the T-R transition. Scheme 5 is a more complete version of Scheme 4 above.

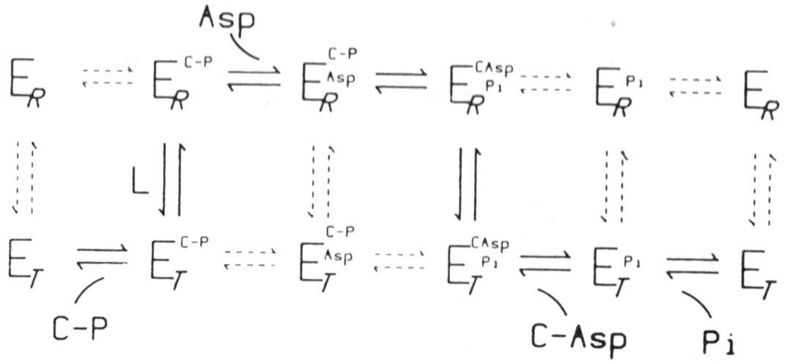

SCHEME 5.   (Reprinted from Hsuanyu, Y. and Wedler, F. C., *J. Biol. Chem.*, 263, 4172, 1988. With permission.)

Systematic kinetics and analytical procedures have been devised recently for using isotope exchanges to distinguish among various possible modes of modifier action.[22] These approaches, with newly devised simulation programs, were used to determine in detail which kinetic steps in the ATCase reaction are altered by ATP and CTP.[23]

Equilibrium isotope-exchange kinetics (EIEK) allow one to observe both the rate-limiting and *faster* steps simultaneously, which obviously can provide unique insights into which steps change in "modified" forms of an enzyme. The data with ATCase are summarized in Scheme 6: effectors differentially alter the binding of Asp, $k_{on} > k_{off}$, rather than having a direct effect on the T-R transition step. Clearly, since substrate binding and the T-R transition occur as sequential, reversible equilibria, these steps are coupled to each other, so that unequal perturbations in the rates for Asp binding and release will also perturb the *equilibrium* for the adjacent T-R step without any alteration in its *rate,* as is observed experimentally.

Finally, EIEK studies were performed with ATCase that had been modified by site-specific mutation of Tyr-240 → Phe in the c-chain. For wild-type enzyme, a hydrogen bond between Tyr-240 and Asp-271 helps stabilize the T-state; hence, disruption of this bond should perturb the T-R equilibrium toward the R-state. In fact, the Tyr-240 → Phe mutation alters the kinetic mechanism on both sides of the reaction, the rate-limiting step, and the regulatory behavior of ATCase.[24]

ACKNOWLEDGMENTS

I am grateful to Myron Bender for his wit and fortitude and the encouragement provided me over the years, plus other Northwestern University faculty in chemistry and biology who helped me appreciate the value of originality and careful thought. Support for this research was provided by grants from NIH, NSF, the American Cancer Society, and NASA.

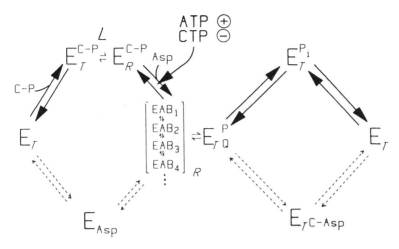

SCHEME 6.   (Reprinted from Hsuanyu, Y. and Wedler, F. C., *J. Biol. Chem.*, 263, 4172, 1988. With permission.)

REFERENCES

1. **Oppenheimer, H. L., Labouesse, B., and Hess, G. P.**, *J. Biol. Chem.*, 241, 2720, 1966.
2. **Matthews, B. W., Sigler, P. B., Henderson, R., and Blow, D. M.**, *Nature*, 214, 652, 1967.
3. **O'Brien, R. D.**, *Toxic Phosphate Esters*, Academic Press, New York, 1960, 106.
4. **Wedler, F. C. and Bender, M. L.**, *J. Am. Chem. Soc.*, 91, 3894, 1969.
5. **Wedler, F. C. and Bender, M. L.**, *J. Am. Chem. Soc.*, 94, 2101, 1972.
6. **Wedler, F. C., Uretsky, L. S., McClune, G.**, *Biochim. Biophys. Acta*, 370, 541, 1974.
7. **Wedler, F. C., Uretsky, L. S., McClune, G., and Cencula, J.**, *Arch. Biochem. Biophys.*, 170, 476, 1975.
8. **Stadtman, E. R. and Ginsburg, A.**, in *The Enzymes*, Vol. 10, 3rd ed., Boyer, P. D., Ed., Academic Press, New York, 1974, 755.
9. **Rhee, S. G., Chock, P. B., Wedler, F. C., and Sugiyama, Y.**, *J. Biol. Chem.*, 256, 644, 1981.
10. **Meek, T. D. and Villafranca, J. J.**, *Biochemistry*, 19, 5513, 1981.
11. **Wedler, F. C., Sugiyama, Y., and Fisher, K. E.**, *Biochemistry*, 21, 2168, 1982.
12. **Wedler, F. C. and Boyer, P. D.**, *J. Biol. Chem.*, 247, 984, 1972.
13. **Wedler, F. C.**, *J. Biol. Chem.*, 249, 5080, 1974.
14. **Wedler, F. C., Shreve, D. S., Kenny, R. M., Ashour, A. E., Carfi, J., and Rhee, S. G.**, *J. Biol. Chem.*, 255, 9507, 1980.
15. **Wedler, F. C., Shreve, D. S., Fisher, K. E., and Merkler, D. J.**, *Arch. Biochem. Biophys.*, 211, 276, 1981.
16. **Merkler, D. J., Srikumar, K., and Wedler, F. C.**, *Biochemistry*, 26, 7805, 1987.
17. **Bild, G. S., Boyer, P. D., and Kohlbrenner, W. E.**, *Biochemistry*, 19, 5774, 1980.
18. **Kantrowitz, E. R., Pastra-Landis, S. C., and Lipscomb, W. N.**, *Trends Biochem. Sci.*, 5, 124, 1980.

19. **Silverstein, E.,** 160th ACS Natl. Meet. Biol. Abstr. 49, 1970.
20. **Wedler, F. C. and Gasser, F. J.,** *Arch. Biochem. Biophys.,* 163, 57, 1974.
21. **Stark, G. R.,** *J. Biol. Chem.,* 246, 3064, 1971.
22. **Wedler, F. C. and Shalongo, W. H.,** *Methods Enzymol.,* 87, 647, 1982.
23. **Hsuanyu, Y. and Wedler, F. C.,** *J. Biol. Chem.,* 263, 4172, 1988.
24. **Hsuanyu, Y., Wedler, F. C., Kantrowitz, E. R., and Middleton, S. A.,** *J. Biol. Chem.,* 264, 17259, 1989.

Chapter 13

EFFECT OF GLYCOSYLATION ON THE ENZYMATIC ACTIVITY OF TISSUE-TYPE PLASMINOGEN ACTIVATORS

Arthur J. Wittwer, Susan C. Howard, and Joseph Feder

TABLE OF CONTENTS

I. INTRODUCTION

Human tissue-type plasminogen activator (tPA) is a 64- to 67-kDa glycosylated serine protease. It catalyzes cleavage of a single peptide bond in the abundant plasma zymogen, plasminogen, thus activating it to the fibrinolytic protease, plasmin (see reviews in References 1, 10, and 19). Plasminogen activation by tPA is stimulated by fibrin, and both tPA and plasminogen bind to fibrin with moderate ($K_d = 0.1$ to 10 μM) affinity (reviewed in Reference 13). Endogenous plasma tPA is responsible for the gradual, spontaneous lysis of fibrin blood clots.[60] Therapeutic doses of tPA, produced by recombinant DNA technology, can cause the rapid lysis of coronary fibrin clots in patients with myocardial infarction (see reviews in References 6 and 50).

A. STRUCTURE AND FUNCTION

Because of its importance in hemostasis, many investigators have sought to understand the structural factors important in tPA activity. These studies have focused on the domain structure of tPA and the sites of N-linked glycosylation (Figure 1). Plasmin can cleave native single-chain tPA (sc-tPA) to a disulfide-linked two-chain species (tc-tPA).[43] Although this cleavage is analogous to that which occurs in zymogen activation, sc-tPA is quite active,[39,45] and the physiological consequences of this cleavage are not yet understood. Five disulfide-bonded structural domains are present in tPA which are homologous to (1) the finger repeats in fibronectin, (2) epidermal growth factor, (3, 4) tandem "kringle" structures, and (5) serine proteases.[2,35] N-linked glycosylation can occur on three of these domains: the first kringle (K1) at Asn-117, the second kringle (K2) at Asn-184, and the serine protease domain (SP) at Asn-448;[3,35] these will be referred to as the K1, K2, and SP sites, respectively. Glycosylation at the K2 site is variable, and natural glycosylation variants result from the presence (type I) or absence (type II) of oligosaccharide at K2 (Figure 1).[37] Structure-function studies have indicated that the finger (F) and K2 domains are important in fibrin binding and stimulation of the plasminogen activation performed by the SP domain.[51] The F domain appears to interact with native fibrin, while K2 has a lysine binding site thought to interact with fibrin partially degraded by plasmin.[11,52,53] The growth factor (G) and K1 domains seem to play a role in the clearance of tPA from the circulation via protein- and carbohydrate-mediated mechanisms, respectively.[7,20,23]

B. ACTIVITY ASSAYS

To correctly describe the effect of structure on tPA function, it is critical that the enzymatic assay be sensitive to the variables examined. In addition to its natural substrate, plasminogen, tPA can also hydrolyze chromogenic peptide substrates.[41,43,55] Thus, tPA can be assayed using a direct amidolytic assay (Figure 2a), or indirectly, employing plasminogen and either an artificial

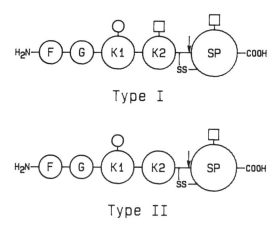

Type I

Type II

FIGURE 1. Schematic of the tPA molecule showing the type I and type II glycoforms. The N terminus (H_2N-) and C terminus ($-COOH$) are indicated. Structural domains are represented by circles: F, fibronectin finger; G, growth factor; K1, first kringle; K2, second kringle; SP, serine protease. Glycosylation sites are indicated by a small circle on K1 (site Asn-117), representing predominantly high-mannose oligosaccharide, and squares on K2 (site Asn-184) and SP (site Asn-448), representing predominantly complex oligosaccharides. An arrow shows the point at which plasmin cleaves sc-tPA to form a disulfide-linked, two-chain species.

or natural plasmin substrate (Figure 2b to d). The indirect amidolytic assay can be unstimulated (Figure 2b), or can include fibrin or fibrin-like stimulators (Figure 2c), which increase the rate of plasmin generation 100- to 1000-fold.[39,58,61] The chromogenic plasmin substrate used in such assays allows the reaction to be monitored continuously and kinetic information obtained. In the stimulated assay (Figure 2c), this reveals a lag phase followed by a more rapid reaction rate, during which activity is typically measured. The increased reaction rate following this lag is due to the action of plasmin on the fibrin or fibrinogen fragment stimulator and the generation of more potent stimulatory species.[4,30] Fibrin can be used as the plasmin substrate in an indirect assay, typified by the clot lysis (Figure 2d) or fibrin plate assays.[12,18] In these assays, tPA is either incorporated into a plasminogen-containing fibrin clot or applied to the surface of a plasminogen-containing fibrin gel. Fibrin degradation is determined by the time taken for a glass bead to pass through the clot or by measurement of the zone of clearing on the plate. Such end-point assays cannot be monitored continuously and so do not allow the examination of kinetics or phases. Understandably, the effect of glycosylation on tPA activity may depend on the type of assay employed to measure it. Glycosylation may affect the cleavage of plasminogen, but not the hydrolysis of a

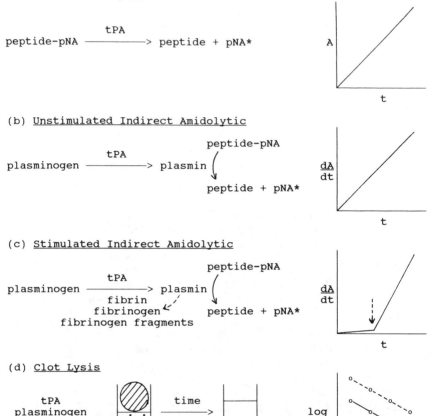

FIGURE 2.   Common assays for tPA. (a) Direct amidolytic assay; (b) unstimulated indirect amidolytic assay; (c) stimulated indirect amidolytic assay; (d) clot lysis assay. For each assay, the reaction scheme is indicated, along with a representation of the data plot. Release of the *para*-nitroaniline chromophore (pNA*) from a *para*-nitroanilide substrate (peptide-pNA) is plotted as absorbance (A) vs. time (t) in (a), and as first-derivative plots (dA/dt vs. t) (b) and (c). Assay conditions are chosen so that these y-axis units are proportional to the pNA* concentration in (a) and plasmin concentration in (b) and (c); thus, the plots show product formation as a function of time. The dashed arrows in (c) indicate partial plasmin degradation of the fibrin(ogen) stimulator, generating more potent stimulatory species which increase the reaction rate. The two linear plots in (d) represent two tPA samples, one (dashed line) less active than the other (solid line). See text for additional details.

low-molecular-weight peptide. One will not see an effect on plasminogen- or fibrin-dependent activity unless these are included, and only a kinetic assay allows the most rapid phase of a fibrin-stimulated reaction to be examined in isolation.

C. GLYCOSYLATION HETEROGENEITY

A search for an effect of glycosylation on tPA activity was inspired by the remarkable variety of carbohydrate structures found both within and between different cell sources of tPA. Type I and type II glycoforms have been separated by lysine-Sepharose chromatography of native human tPA from cultured Bowes melanoma cells (m-tPA),[12,44] human colon fibroblasts (hcf-tPA),[33] and in recombinant human tPA produced in murine C127 cells (C127 r-tPA)[34] and Chinese hamster ovary cells (CHO r-tPA).[34,61] In addition to site occupancy, another source of heterogeneity is the large number of different carbohydrate structures present at each site. Analyses of m-tPA,[33,38] hcf-tPA,[33] C127 r-tPA,[34,36] and CHO r-tPA[8,29,34,46,47] have shown that primarily high-mannose oligosaccharides are found on K1 (Asn-117), while mostly complex-type oligosaccharides are found on K2 (Asn-184) and SP (Asn-448). Three of the more common sugar structures identified on hcf-tPA and m-tPA are shown in Figure 3, and illustrate the distinction between high-mannose and complex-type structures. At each glycosylation site, however, many different additional structures have been found, and a different repertoire of specific structures are found in tPA from each cell source. For example, from the data of Parekh et al.,[33] it can be calculated that there are well over 400 natural glycosylation variants of type I hcf-tPA, over 40 variants of type II hcf-tPA, over 200 variants of type I m-tPA, and over 20 variants of type II m-tPA. Even with this variety, however, hcf- and m-tPA are structurally distinct from each other, i.e., there are no identically glycosylated molecules made by the two cell types.[33] CHO- and C127 r-tPA also show distinct patterns of carbohydrate structures typical of cell type-specific glycosylation.[34,36,46]

From these considerations, we and others have examined two types of differences between naturally glycosylated tPA which could affect activity. These are site *occupancy*, typified by the distinction between types I and II, and site *composition*, exemplified by type I or II species from different cell sources. The effect of site Asn-184 occupation can be examined by comparing the activity of type I and type II from the same cell source. The effect of the type of oligosaccharide at a given site, however, is examined by comparing the activities of type I or type II tPAs from different cell sources. To these we can also add comparisons made with unnatural tPA molecules created by mutation, inhibitors, exoglycosidases, or endoglycosidases, which have intentionally altered site occupancy or composition.

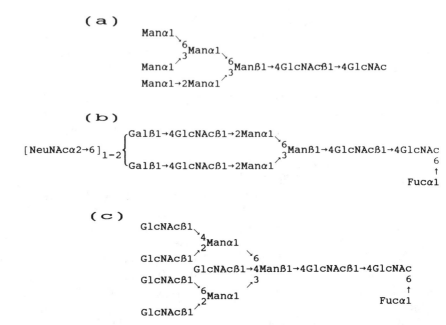

FIGURE 3.   Representative oligosaccharide structures found on hcf-tPA and m-tPA. (a) High-mannose structure containing six mannose residues found on both hcf-tPA and m-tPA; (b) biantennary complex oligosaccharide with core fucosylation and variable sialylation found only on hcf-tPA; (c) tetraantennary complex oligosaccharide with core fucosylation, bisecting GlcNAc, and lacking terminal galactose. This structure is found only on m-tPA. Structures and symbolic notation are as described by Parekh et al.[33]

II. EFFECT OF SITE OCCUPANCY ON ACTIVITY

A. NATURALLY OCCURRING tPA GLYCOFORMS

Type I and II tPA can be separated by chromatography on lysine-Sepharose (Figure 4). The type II variant, which lacks glycosylation on K2, interacts more strongly with this matrix. Although these glycoforms have equal direct amidolytic activity,[41,58,59] the type II variant from m-tPA or hcf-tPA was found to have 23 to 50% greater activity in fibrin clot lysis[12,58] and a two-fold or greater activity in a fibrinogen fragment-stimulated indirect amidolytic assay[58] (Figure 4). The same activity difference holds for type I and II r-tPA from CHO and C127 expression systems[21,34,62] (Figure 4). Interestingly, domain deletion variants of C127 r-tPA which lack F, G, or K1 still appear to produce type I and II glycoforms, due to the presence or absence of K2 glycosylation.[56] When a variant containing only the K1, K2, and SP domains (K1 + K2 + SP) was separated on lysine-Sepharose, the type II form had a stimulated indirect amidolytic activity eight times that of the type I form, and a fibrin plate activity which was 2.2 times greater.[57]

The indirect amidolytic assay has been used to determine the kinetics of plasminogen activation by type I and type II tPA. Although Zamarron et al.[61] concluded that there was no difference in the kinetic parameters of types I and II CHO r-tPA, this study compared equal fibrin-stimulated activities of the two forms, rather than equal amounts of tPA protein. This method of determining enzyme concentration would decrease or eliminate any activity differences due to catalytic efficiency. Using a fibrinogen fragment-stimulated indirect amidolytic assay where tPA protein was determined by antigen measurements, the type II species from hcf- and m-tPA were found to have a k_{cat} 2.5 times that of type I species with approximately equal K_ms for plasminogen.[58] In the absence of stimulator, types I and II tPA showed equal activity in this assay, suggesting that K2 glycosylation does not affect basal plasminogen activation, but instead inhibits stimulation of this activity by fibrin or fibrinogen fragments.[58]

The activity difference between types I and II tPA is typically greater when measured in the stimulated indirect amidolytic assay than when measured in a clot lysis or fibrin plate assay. Activity in the stimulated indirect assay is measured after partial plasmin degradation of the stimulator (Figure 2c), a process which yields C-terminal lysine residues thought to interact with the lysine binding site on K2.[4,11] Glycosylation on K2 appears to partially mask the K2 lysine binding site, reducing lysine affinity and decreasing activity in the stimulated indirect assay (Figure 4). Because stimulation by native stimulator (mediated by the F domain in addition to K2) plays a more important role in the clot lysis or fibrin plate assays, these are less sensitive to the activity difference between type I and type II.

B. OTHER tPA SPECIES WITH REDUCED LEVELS OF OLIGOSACCHARIDE

Increases in tPA activity have also been observed when glycosylation site occupancy has purposely been decreased by treatment of cells with tunicamycin, removal of oligosaccharides with N-glycanase, or site-directed mutagenesis. An early study suggested that m-tPA isolated from cells treated with tunicamycin, an inhibitor of N-linked glycosylation, was unaltered in its fibrin binding or fibrin-stimulated activity.[44] However, increased fibrin binding and clot lysis activity was found for carbohydrate-depleted CHO r-tPA from tunicamycin-treated cultures.[16] Hcf- and m-tPA from tunicamycin-treated cells had an increased stimulated indirect amidolytic activity similar to that of a type II variant, while the unstimulated indirect activity increased two- to fourfold over that seen for either type I or type II.[58] Similar results were reported for C127 r-tPA.[57] Tunicamycin treatment increased activity about twofold in both the stimulated and unstimulated indirect amidolytic assay, and increased fibrin plate activity by 50%. Wilhelm et al.[57] also showed that the fibrin plate activity of a variant of tPA which lacked the F and G domains (K1 + K2 + SP) increased 4.6-fold when produced in the presence of tunicamycin.

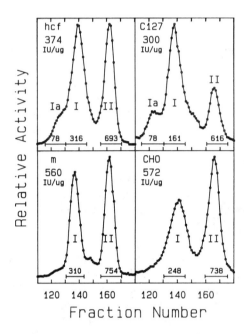

FIGURE 4.   Lysine-Sepharose chromatography of 2 mg of hcf-tPA ("hcf"), 4 mg of m-tPA ("m"), 0.30 mg of C127 r-tPA ("C127"), and 0.37 mg of CHO r-tPA ("CHO"). The specific activity of the unfractionated tPAs in the stimulated indirect amidolytic assay is given in the top-right corner of each panel. Fractions were monitored by measuring the rate of direct hydrolysis of S-2322 (H-D-Val-Gly-Arg-*p*-nitroanilide, KabiVitrum) and the appearance of 405-nm absorbance to yield relative activities. The y-axis scales are as follows: (hcf), −2 to 39 mAU/min; (m), −5 to 108 mAU/min; (C127), −0.06 to 1.21 mAU/min; (CHO), −0.3 to 6.2 mAU/min. Fractions pooled for activity measurement are indicated by bars, and the specific activity of these pools in the stimulated indirect amidolytic assay is given above each bar in IU/μg. Pools are designated as follows: "Ia", shoulder to the type I peak; "I", type I; "II", type II. Source of tPA, chromatography, and enzymatic assays were as described previously.[34,58]

The enzyme *N*-glycanase, which removes accessible N-linked oligosaccharides and leaves an Asp residue instead of Asn on the protein, has also been used to generate less-glycosylated tPA species. In this manner, Wilhelm et al.[57] obtained partially deglycosylated forms of C127 r-tPA and of a type I C127 K1 + K2 + SP deletion variant. The fibrin plate activity of these carbohydrate-depleted species increased 34% and fourfold over control, respectively. In a different study, *N*-glycanase treatment of a deletion mutant containing only the K2 and SP domains increased activity 2.1-fold in the unstimulated indirect assay, 4.7-fold in the stimulated indirect assay, and 3.5-fold in the clot lysis assay.[31]

Finally, several reports have described the activity of r-tPA species in which N-linked glycosylation was prevented by site-directed mutagenesis of Asn in the glycosylation sites to Gln. Deletion of the K1 (Asn-117) site from a

K1 + K2 + SP variant expressed in CHO cells increased clot lysis activity 33%, while deleting all three sites increased activity 88%.[16] Other investigators found no consistent increase or decrease in activity over native tPA when the SP (Asn-448) site was removed from C127 r-tPA[25] or when the K1 (Asn-117), K2 (Asn-184), or SP sites were individually deleted from CHO r-tPA.[15] Results from the latter study appear to be inconsistent with the increased activity observed for type II tPA, since deletion of the K2 site should yield a type II-like molecule. However, native CHO r-tPA already has a high activity because of a high proportion of type II glycoforms (Figure 4), making any increase due to K2 site deletion less prominent. It would have been more informative to compare such site deletion variants to the fully glycosylated enzyme, type I CHO r-tPA. Significant increases in activity over native CHO r-tPA, however, were seen by Haigwood et al.[15] when either *both* the K1 and K2 sites, or all three sites together, were deleted. Both clot lysis and stimulated indirect amidolytic activity were increased two- to threefold. Unstimulated indirect activity was increased about twofold in a mutant lacking both the K1 and K2 sites, but was increased an impressive sevenfold when the K1, K2, and SP sites were all deleted in the same molecule. These results, along with those previously described, suggest a particularly important role of the SP oligosaccharide in inhibiting plasminogen activation in the absence of fibrin.

D. SUMMARY OF SITE OCCUPANCY EFFECTS

The above studies suggest that either the natural absence or the purposeful removal of oligosaccharide from tPA tends to increase its activity. The magnitude of the increase seems to depend both on the particular glycosylation site involved and the assay used to measure it. It should be noted that no difference in direct amidolytic activity was ever observed in these studies. The presence or absence of carbohydrate does not affect the hydrolysis of a low molecular weight peptide, but instead influences interaction with the natural macromolecular substrate and effector molecules.

The effect of glycosylation site occupancy on tPA activity in the indirect amidolytic assay in the presence and absence of stimulator is summarized in Figure 5. Comparison of type I ("all 3") and type II ("K1 + SP") indicates that occupancy of the K2 site depresses stimulated, but not unstimulated, activity. The remaining oligosaccharides at K1 and SP appear important for inhibiting unstimulated activity, since their removal results in a substantial increase. These results are at least partially consistent with the model that K2 oligosaccharide inhibits the ability of the stimulator fragments to bind to the K2 lysine binding site, while the SP oligosaccharide inhibits plasminogen binding. Since improved interaction with plasminogen should also increase stimulated activity, a further increase in stimulated activity above the type II level for tPA from tunicamycin-treated cells ("none" in Figure 5) might be expected. This may not have been observed in these experiments due to the

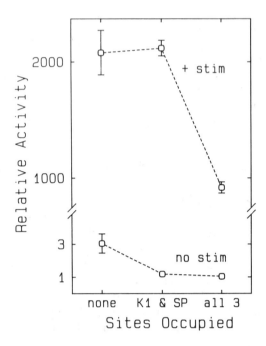

FIGURE 5.   Influence of glycosylation site occupancy
on activity in the stimulated ("+ stim") and unstimulated
("no stim") indirect amidolytic assay. Data presented are
an average of the results previously reported for hcf-tPA
and m-tPA.[58] Error bars indicate the standard error of the
mean. Standard errors less than the half-width of the plot-
ting symbol are not shown. Activities are shown for unglycosylated tPA produced in the presence of tunicamycin
("none"), for type II tPA glycosylated at the K1 (Asn-
177) and SP (Asn-448) sites ("K1 & SP"), and for
type I tPA glycosylated at all three sites ("all 3"). Mannose labeling indicated that the unglycosylated tPA produced in the presence of tunicamycin had an oligosaccharide content 13% of that of control tPA. For clarity,
a dashed line connects points measured in the same type
of assay.

incomplete action of tunicamycin, which reduced mannose incorporation into
tPA by 87%.[58] Note that the extent to which tPA is stimulated by fibrinogen
fragments is greatest for the type II form, but may be least for the unglycosylated enzyme (see also Reference 15). This may imply that totally deglycosylated tPA may be less fibrin specific, a possibility which should be
considered in the design of second-generation tPA variants. Finally, clot lysis
and fibrin plate assay results (not shown in Figure 5) suggest a significant
influence of all three glycosylation sites on fibrin-stimulated plasminogen
activation. Activity increases when the K2 oligosaccharide is absent in type
II, and then increases further when the K1 and SP carbohydrates are removed.

III. EFFECT OF OLIGOSACCHARIDE COMPOSITION ON ACTIVITY

A. NATURALLY OCCURRING tPA VARIANTS

Although some of the variation in native tPA specific activity can be explained by the relative proportions of type I and type II glycoforms (Figure 4), it is evident that tPAs with identical site occupancy can still have significantly different activities. The lower-stimulated indirect amidolytic activity of the type Ia species in C127 r-tPA and hcf-tPA compared to type I are examples of this, as is also the lower activity of type I C127 r-tPA compared to type I from other cell sources[21] (Figure 4). It is also of interest that the generally lower-activity hcf-tPA and C127 r-tPA have a higher proportion of larger, galactosylated tri-, tetra-, and pentaantennary complex oligosaccharides (37 and 36%, respectively) than do the higher-activity m-tPA and CHO r-tPA. These observations indicate that the type of oligosaccharide which occupies a given glycosylation site can influence tPA activity.

B. VARIANTS PRODUCED BY EXOGLYCOSIDASE TREATMENT

Little et al.[26] report the results of treating m-tPA with *endo*-β-*N*-acetyl-glucosaminidase H (*endo*-H), an enzyme which cleaves high-mannose oligosaccharides between the two proximal GlcNAc residues (see Figure 3). Altering the high-mannose sugars in this manner did not affect fibrin plate or stimulated indirect amidolytic activity. Opdenakker et al.[32] treated m-tPA with neuraminidase, β-galactosidase, and α-mannosidase to remove terminal sialic acid, galactose, and mannose residues, respectively. Neuraminidase decreased fibrin plate activity by 53%, β-galactosidase had little effect (19% decrease), and α-mannosidase increased the activity by 55%. No subsequent reports have appeared on the effect of α-mannosidase on tPA activity. However, the effect of neuraminidase has recently been reexamined, and substantial *increases* in activity have been noted when tPA from sources other than Bowes melanoma cells are treated. Table 1 summarizes the data of Wilhelm et al.[57] and unpublished results from this laboratory. Table 1 also indicates that the extent of the activity increase upon neuraminidase treatment tended to correlate with the level of sialic acid in the tPA. The low degree of sialylation of m-tPA may explain why results with this tPA were different. As discussed by Wilhelm et al.,[57] the high degree of sialylation of the K1 + K2 + SP deletion mutant expressed in C127 cells may be one reason for its lowered activity compared to normal tPA. Wilhelm et al.[56] have shown that the K1 glycosylation site, which is normally high mannose, undergoes processing to sialylated, complex structures when the F or G domains have been deleted or altered. High levels of sialylation appear to be one reason for the extremely low activity of the type Ia species separable on lysine-Sepharose (Figure 4, Table 1).

TABLE 1
Effect of Neuraminidase Treatment on tPA Specific Activity

tPA species	Specific activity (IU/μg)		Ratio (treated: control)	Sialic acid (mol/ mol tPA)
	Control	Treated		
m-tPA[a]	900	900	1.0	0.4
m-tPA[b]	500	527	1.1	0.1
hcf-tPA[b]				
Unfractionated	270	583	2.2	2
Type Ia	41	190	4.6	—[c]
Type I	211	450	2.1	—
Type II	559	916	1.6	—
C127 r-tPA[a]	380	470	1.2	2
C127 r-tPA[b]	306	529	1.9	3
CHO r-tPA[b]	439	665	1.5	2
K1 + K2 + SP tPA[a]				
Unfractionated	150	—	—	5
Type I	90	360	4.0	—
Type II	200	—	—	—

[a]   Data from Wilhelm et al. (1990). A fibrin plate assay was used. K1 + K2 + SP tPA was expressed in C127 cells and represents a deletion mutant of tPA lacking the F and G domains. This mutant was separable on lysine-Sepharose into type I and type II glycoforms.

[b]   Data from S. C. Howard and A. J. Wittwer, unpublished observations. The tPA sample at 25 μg/ml was incubated for 16 h at room temperature with or without 28 mU/ml *Arthrobacter ureafaciens* neuraminidase (Calbiochem) in 140 mM sodium acetate, 280 mM ammonium bicarbonate buffer, pH 5.0. Samples were assayed in a stimulated indirect amidolytic assay.[58] The sialic acid content of the indicated tPA species was estimated from oligosaccharide composition data.[33,34]

[c]   Data not given or estimated.

C. VARIANTS PRODUCED BY PROCESSING INHIBITORS

To further examine the influence of oligosaccharide composition on tPA activity, recombinant C127 or CHO cells were treated with the inhibitor deoxymannojirimycin, a golgi α-mannosidase inhibitor which prevents the processing of high-mannose oligosaccharides to complex forms. Cells treated in this manner express tPA with predominantly high-mannose oligosaccharides and which has a stimulated indirect amidolytic activity significantly greater than tPA from untreated cells.[62] The increases in tPA activity which were observed under these conditions were similar to those resulting from neuraminidase treatment (Table 1).

D. SUMMARY OF COMPOSITION EFFECTS

The activity increases seen when tPA is treated with neuraminidase or produced in the presence of deoxymannojirimycin imply that sialylated, complex oligosaccharides depress fibrin-stimulated plasminogen activation by tPA. As was the case with site occupancy, this effect was not seen in any of the above studies when tPA activity was measured with a low molecular weight peptide substrate. This suggests that different oligosaccharide structures can have different effects on the interaction of the enzyme with plasminogen and fibrin.

IV. DISCUSSION

A. EFFECTS OF GLYCOSYLATION ON ACTIVITY: SITE OCCUPANCY AND COMPOSITION

The results described above indicate that both increased glycosylation site occupancy and increased processing to complex, sialylated oligosaccharides serves to decrease tPA activity. Data presented for hcf-tPA glycoforms in Table 1 allow an examination of the relative effects of both site occupancy and composition on stimulated tPA activity. There is a 22-fold increase in activity when control type Ia hcf-tPA (41 IU/µg) is compared to neuraminidase-treated type II hcf-tPA (916 IU/µg). Assuming that sialic acid and site occupancy exert independent effects, this activity difference can be broken down into several components. Since the K1 glycosylation site is completely high mannose and type II lacks oligosaccharide at K2, sialylation can only occur at K2 and SP in type I hcf-tPA, and only at SP in type II hcf-tPA. Thus, neuraminidase treatment of type II hcf-tPA results in a 1.6-fold increase in activity which is entirely attributable to removal of sialic acid from the SP site. Neuraminidase treatment of type I hcf-tPA removes sialic acid from SP and K2 and increases activity 2.1-fold. Since the SP oligosaccharide composition is the same in both type I and type II hcf-tPA,[33] it can be calculated that removal of sialic acid from the K2 site alone would result in a 2.1/1.6 or 1.3-fold increase. The effect of removing sialic acid from K2 can also be determined from the effect of removing a neutral oligosaccharide from K2 (neuraminidase-treated type II compared to neuraminidase-treated type I), which results in a 2.0-fold increase, and the effect of removing a sialylated oligosaccharide from K2 (type II compared to type I), which results in a 2.6-fold increase. Once again, a 2.6/2.0 or 1.3-fold increase is calculated. Table 1 also indicates that type I has a 5.0-fold greater activity than type Ia, an effect which must be due to composition effects. Calculations comparing control and neuraminidase-treated types Ia and I suggest that decreased sialylation at K2 and SP accounts for a portion (2.1-fold) of this increase. The remaining 2.4-fold activity difference is presumably due to differences in the neutral cores of the complex oligosaccharides at SP and K2. Thus, the 22-fold activity difference between type Ia hcf-tPA and neuraminidase-treated type II hcf-tPA can be dissected as follows: (1) a total of 4.4-fold due to sialylation, about half of which results from structures unique to type Ia, (2) 2.0-fold due to K2 occupancy, and (3) 2.4-fold due to oligosaccharide composition differences other than sialylation. These composition differences probably involve highly branched, complex structures on the type Ia molecules which could provide the additional sites for sialylation. In conclusion, site occupancy and composition appear to have additive effects on tPA activity, with composition responsible for the greatest differences between naturally occurring glycosylation variants.

B. MECHANISM OF OLIGOSACCHARIDE INHIBITION OF ACTIVITY

The mechanism whereby oligosaccharides interfere with the binding of plasminogen and fibrin-like simulators to tPA is not understood. A carbohydrate of particular structure could bind to a lectin-like site on the enzyme, substrate, or effector surface, producing an allosteric effect on activity or preventing productive binding. More simply, the approach of large substrate or effector molecules could be inhibited by the charge or bulk of the oligosaccharide, without any highly specific interactions taking place. A typical N-linked carbohydrate moiety can mask a considerable area of a protein's surface.[28] On tPA, the K1 and K2 oligosaccharides are almost as big as the kringles to which they are attached.[33]

C. PHYSIOLOGICAL RELEVANCE

The data presented above suggest that the *in vitro* activity of different naturally produced forms of tPA is inversely correlated with both the extent of glycosylation and the extent of processing to complex, sialylated structures on the enzyme. In regard to the potential physiological relevance of these observations, several cautions are in order. First, the glycosylation pattern of tPA as it is naturally present in human plasma or tissue is unknown. Second, it has not yet been demonstrated that the differently glycosylated forms of tPA possess different activities in any *in vivo* system. If, however, the tenfold range of specific activities seen across lysine-Sepharose profiles of tPA isolated from *in vitro* cell culture is representative of tPA populations *in vivo*, a physiological role which exploits this variety is implied.

In vivo thrombolysis involves a complex interplay between tPA, plasminogen, specific inhibitors, and fibrin. sc-tPA is at the top of the fibrinolytic cascade. Initial activation of plasminogen by sc-tPA will generate plasmin. Plasmin, in turn, can convert sc-tPA to tc-tPA, which may be more active under physiological conditions,[5,54] but may also react more rapidly with natural inhibitors.[9,22,24,40] It has recently been shown that the conversion of sc-tPA to tc-tPA, like stimulated tPA activity, is inhibited about twofold by glycosylation on K2, i.e., type II sc-tPA undergoes this conversion twice as fast as type I sc-tPA.[59] In addition, plasmin initially generated in fibrinolysis can convert native plasminogen to more readily activatable forms[49] and can degrade fibrin to produce more potent stimulatory species.[30,48] Thus, even a small difference in initial plasmin generation, as a result of glycosylation effects on sc- and tc-tPA activity, could be magnified many fold in terms of final plasmin levels and fibrin dissolution when the accelerating effect of these feedback events are considered. Since a heavily glycosylated type I sc-tPA is less active, but potentially more resistant to conversion to tc-tPA and reaction with inhibitor, such forms could represent "slow", but persistent tPA species. A less modified, more active type II sc-tPA could be "fast", but less persistent. As previously discussed,[58] the diversity of tPA forms may allow for the balance in fibrinolytic activity which is required for proper wound healing, thus providing for initial clot persistence, but eventual dissolution.

Finally, there is evidence that the glycosylation pattern of tPA made by a given cell can be regulated by exogenous factors. We have observed that when expression of hcf-tPA is increased by the addition of growth factors, or when production of C127 r-tPA is increased by butyrate,[27] the relative amount of type II tPA increases, and the low-activity type Ia lysine-Sepharose species disappears.[62] As expected, these changes resulted in increased activity. It is possible that physiologically relevant factors, which serve to increase tPA levels during wound healing, tissue remodeling, or neoplastic transformation,[14,17,42] may also result in a less processed, less glycosylated tPA having enhanced activity.

ACKNOWLEDGMENTS

The authors gratefully acknowledge the following co-workers at the Monsanto Company (St. Louis, MO): David C. Tiemeier, T. V. Ramabhadran, Beverly A. Reitz, M. Brenda Hebert, Thomas G. Warren, Mark O. Palmier, and T.-C. Wun for providing CHO and C127 r-tPA used in some of the unpublished experiments described herein; Linda S. Carr and Deena Reyes for expert technical assistance; Charles Lewis, Gene Pegg, and Lani Patton for cell culture support; and Joseph K. Welply for oligosaccharide analyses.

REFERENCES

1. **Bachmann, F. and Kruithof, E. K. O.**, *Semin. Thromb. Hemostasis*, 10, 6, 1984.
2. **Banyai, L., Varadi, A., and Patthy, L.**, *FEBS Lett.*, 163, 37, 1983.
3. **Bennett, W. F.**, *Thromb. Haemostasis*, 50 (Abstr.), 106, 1983.
4. **Beckmann, R., Geiger, M., and Binder, B. R.**, *J. Biol. Chem.*, 263, 7176, 1988.
5. **Boose, J. A., Kuismanen, E., Gerard, R., Sambrook, J., and Gething, M.-J.**, *Biochemistry*, 28, 635, 1989.
6. **Braunwald, E.**, *J. Am. Coll. Cardiol.*, 12, 85A, 1988.
7. **Browne, M. J., Carey, J. E., Chapman, C. G., Tyrrell, A. W. R., Entwisle, C., Lawrence, G. M. P., Reavy, B., Dodd, I., Esmial, A., and Robinson, J. H.**, *J. Biol. Chem.*, 263, 1599, 1988.
8. **Carr, S. A., Roberts, G. D., Jurewicz, A., and Frederick, B.**, *Biochimie*, 70, 1445, 1988.

9. Chmielewska, J., Rånby, M., and Wiman, B., *Biochem. J.*, 251, 327, 1988.
10. Collen, D., *Thromb. Res.*, Suppl. 8, 3, 1988.
11. de Vries, C., Veerman, H., and Pannekoek, H., *J. Biol. Chem.*, 264, 12604, 1989.
12. Einarsson, M., Brandt, J., and Kaplan, L., *Biochim. Biophys. Acta*, 830, 1, 1985.
13. Fears, R., *Biochem. J.*, 261, 313, 1989.
14. Gerard, R. D. and Meidell, R. S., *Annu. Rev. Physiol.*, 51, 245, 1989.
15. Haigwood, N. L., Mullenbach, G. T., Moore, G. K., DesJardin, L. E., Tabrizi, A., Brown-Shimer, S. L., Stauß, H., Stöhr, H. A., and Pâques, E.-P., *Protein Eng.*, 2, 611, 1989.
16. Hansen, L., Blue, Y., Barone, K., Collen, D., and Larsen, G. R., *J. Biol. Chem.*, 263, 15713, 1988.
17. Hart, D. A. and Rehemtulla, A., *Comp. Biochem. Physiol. B*, 90, 691, 1988.
18. Haverkate, F. and Brakman, P., *Prog. Chem. Fibrinolysis Thrombolysis*, 1, 151, 1975.
19. Higgins, D. L. and Bennett, W. F., *Annu. Rev. Pharmacol. Toxicol.*, 30, 90, 1990.
20. Hotchkiss, A., Refino, C. J., Leonard, C. K., O'Connor, J. V., Crowley, C., McCabe, J., Tate, K., Nakamura, G., Powers, D., Levinson, A., Mohler, M., and Spellman, M. W., *Thromb. Haemostasis*, 60, 255, 1988.
21. Howard, S. C., Wittwer, A. J., Carr, L. S., Harakas, N. K., and Feder, J., *J. Cell Biol.*, 107, (Abstr.), 584, 1988.
22. Hekman, C. M. and Loskutoff, D. J., *Arch. Biochem. Biophys.*, 262, 199, 1988.
23. Kalyan, N. K., Lee, S. G., Wilhelm, J., Fu, K. P., Hum, W.-T., Rappaport, R., Hartzell, R. W., Urbano, C., and Hung, P. P., *J. Biol. Chem.*, 263, 3971, 1988.
24. Korninger, C. and Collen, D., *Thromb. Haemostasis*, 46, 662, 1981.
25. Lau, D., Kuzma, G., Wei, C.-M., Livingston, D. J., and Hsiung, N., *Bio/Technology*, 5, 953, 1987.
26. Little, S. P., Bang, N. U., Harms, C. S., Marks, C. A., and Mattler, L. E., *Biochemistry*, 23, 6191, 1984.
27. Mroczkowski, B., Reich, M., Wittaker, J., Bell, G., and Cohen, S., *Proc. Natl. Acad. Sci. U.S.A.*, 85, 126, 1988.
28. Neuberger, A. and van Deenen, L. L. M., in *Comprehensive Biochemistry*, Vol. 19B, Elsevier, New York, 1982, part 2, 82.
29. Nimtz, M., Noll, G., Pâques, E.-P., and Conradt, H. S., *FEBS Lett.*, 271, 14, 1990.
30. Norrman, B., Wallén, P., and Rånby, M., *Eur. J. Biochem.*, 149, 193, 1985.
31. Obukowicz, M. G., Gustafson, M. E., Junger, K. D., Leimgruber, R. M., Wittwer, A. J., Wun, T.-C., Warren, T. G., Bishop, B. F., Mathis, K. J., McPherson, D. T., Siegel, N. R., Jennings, M. G., Brightwell, B. B., Diaz-Collier, J. A., Bell, L. D., Craik, C. S., and Tacon, W. C., *Biochemistry*, 29, 9737, 1990.
32. Opdenakker, G., Damme, J. V., Bosman, F., Billiau, A., and Somer, P. D., *Proc. Soc. Exp. Biol. Med.*, 182, 248, 1986.
33. Parekh, R. B., Dweck, R. A., Thomas, J. R., Opendakker, G., Rademacher, T. W., Wittwer, A. J., Howard, S. C., Nelson, R., Siegel, N. R., Jennings, M. G., Harakas, N. K., and Feder, J., *Biochemistry*, 28, 7644, 1989.
34. Parekh, R. B., Dwek, R. A., Rudd, P. M., Thomas, J. R., Rademacher, T. W., Warren, T., Wun, T.-C., Hebert, B., Reitz, B., Palmier, M., and Tiemeier, D. C., *Biochemistry*, 28, 7670, 1989.
35. Pennica, D., Holmes, W. E., Kohr, W. J., Harkins, R. M., Vehar, G. A., Ward, C. A., Bennett, W. F., Yelverton, E., Seeburg, P. H., Heyneker, H. L., Goedell, D. V., and Collen, D., *Nature*, 301, 214, 1983.
36. Pfeiffer, G., Schmidt, M., Strube, K.-H., and Geyer, R., *Eur. J. Biochem.*, 186, 273, 1989.
37. Pohl, G., Kallstrom, M., Bergsdorf, N., Wallén, P., and Jörnvall, H., *Biochemistry*, 23, 3701, 1984.

38. **Pohl, G., Kenne, L., Nilsson, B., and Einarsson, M.,** *Eur. J. Biochem.,* 170, 69, 1987.
39. **Rånby, M.,** *Biochim. Biophys. Acta,* 704, 461, 1982.
40. **Rånby, M., Bergsdorf, N., and Nilsson, T.,** *Thromb. Res.,* 27, 175, 1982.
41. **Rånby, M., Bergsdorf, N., Pohl, G., and Wallén, P.,** *FEBS Lett.,* 146, 289, 1982.
42. **Reich, E.,** in *Molecular Basis of Biological Degradative Processes,* Berlin, R. D., Herrmen, H., Leopow, I., and Tanzer, J., Eds., Academic Press, New York, 1978, 155.
43. **Rijken, D. C. and Collen, D.,** *J. Biol. Chem.,* 256, 7035, 1981.
44. **Rijken, D. C., Emeis, J. J., and Gerwig, G. J.,** *Thromb. Haemostasis,* 54, 788, 1985.
45. **Rijken, D. C., Hoylaerts, M., and Collen, D.,** *J. Biol. Chem.,* 257, 2920, 1982.
46. **Spellman, M. W., Basa, L. J., Leonard, C. K., Chakel, J. A., O'Connor, J. V., Wilson, S., and van Halbeek, H.,** *J. Biol. Chem.,* 264, 14100, 1989.
47. **Spellman, M. W.,** *Anal. Chem.,* 62, 1714, 1990.
48. **Suenson, E., Lützen, O., and Thorsen, S.,** *Eur. J. Biochem.,* 140, 513, 1984.
49. **Suenson, E. and Thorsen, S.,** *Biochemistry,* 27, 2435, 1988.
50. **Tiefenbrunn, A. J. and Sobel, B. E.,** *Fibrinolysis,* 3, 1, 1989.
51. **van Zonneveld, A.-J., Veerman, H., and Pannekoek, H.,** *Proc. Natl. Acad. Sci. U.S.A.,* 83, 4670, 1986.
52. **van Zonneveld, A.-J., Veerman, H., and Pannekoek, H.,** *J. Biol. Chem.,* 261, 14214, 1986.
53. **Verheijen, J. H., Caspers, M. P. M., Chang, G. T. G., de Munk, G. A. W., Poawels, P. H., and Enger-Valk, B. E.,** *EMBO J.,* 5, 3525, 1986.
54. **Wallén, P., Pohl, G., Bergsdorf, N., Rånby, M., Ny, T., and Jörnvall, H.,** *Eur. J. Biochem.,* 132, 681, 1983.
55. **Wallén, P., Rånby, M., Bergsdorf, N., and Kok, P.,** *Prog. Fibrinolysis,* 5, 16, 1981.
56. **Wilhelm, J., Lee, S. G., Kalyan, N. K., Cheng, S. M., Wiener, F., Pierzchala, W., and Hung, P. P.,** *Biotechnology,* 8, 321, 1990.
57. **Wilhelm, J., Kalyan, N. K., Lee, S. G., Hum, W.-T., Rappaport, R., and Hung, P. P.,** *Thromb. Haemostasis,* 63, 464, 1990.
58. **Wittwer, A. J., Howard, S. C., Carr, L. S., Harakas, N. K., Feder, J., Parekh, R. B., Rudd, P. M., Dwek, R. A., and Rademacher, T. W.,** *Biochemistry,* 28, 7662, 1989.
59. **Wittwer, A. J. and Howard, S. C.,** *Biochemistry,* 29, 4175, 1990.
60. **Wun, T.-C. and Capuano, A.,** *J. Biol. Chem.,* 260, 5061, 1985.
61. **Zamarron, C., Lijnen, H. R., and Collen, D.,** *J. Biol. Chem.,* 259, 2080, 1984.
62. **Howard, S. C., Wittwer, A. J., and Welply, J. K.,** *Glycobiology,* in press.

Chapter 14

VIOLOGEN MEDIATORS OF *DESULFOVIBRIO DESULFURICANS* HYDROGENASE: STRUCTURE-FUNCTION RELATIONSHIPS

E. Ziomek, W. G. Martin, and R. E. Williams

TABLE OF CONTENTS

I. INTRODUCTION

Hydrogenase (EC 1.12.2.1), the enzyme catalyzing the interconversion of protons and hydrogen molecules, has been isolated from many different sources.[1] The substantial interest in this enzyme and the important role it plays in the energy metabolism and bioenergetics of many chemotropic and phototropic organisms has been reviewed.[1-3]

These recent reviews have emphasized the suitability of the enzyme as a model system for the study of proteins involved in the electron transfer systems of these types of organisms. An additional benefit also has been pointed out: the small size of the genome encoding these proteins allows for relatively easy access to genes encoding the proteins and enzymes which relay energy along the electron transfer chain, thereby facilitating the study of their organization and control.[2]

The reaction catalyzed by the enzyme is as follows:

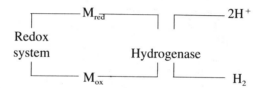

where the redox potential energy (2e), emanating from either chemical or photochemical systems, is intercepted by a mediator (M) and finally transferred, via the intermediacy of the enzyme hydrogenase, to the production of molecular hydrogen from protons. The enzyme-catalyzed reaction is reversible and uses either ferredoxin or various cytochromes as naturally occurring mediators. The difficulties of isolating and purifying the natural systems as well as their instability led Peck and Gest to substitute them with the synthetic compound, methyl viologen (N,N'-dimethyl-4,4'-dipyridine (Table 1, $R = R' = -CH_3$), a low-redox-potential compound which has been previously used as a redox indicator.[4] Subsequent studies have led others to study the use of dibenzyl viologen (Table 1, $R = R' = -CH_2C_6H_5$)[5-7] and several structural variants[8,9] in similar reaction systems. In an attempt to better define the structure-function relationships in the synthetic mediators, a wider range of structural variations of the viologen mediators were tested for their ability to mediate both the hydrogen evolution and uptake reactions catalyzed by the periplasmic hydrogenase isolated from the sulfate-reducing anaerobe, *Desulfovibrio desulfuricans*.

II. EXPERIMENTAL PROCEDURES

A. CHEMICALS

All chemicals used for buffer and media preparation were reagent grade and used as received. Bovine serum albumin (BSA) was purchased for Sigma

TABLE 1
Extinction Coefficients, Redox Potentials, and Hydrogenase Mediator Activity of the One-Electron Reduction Product of the Various Side Chain Modified Viologens

Viologen structure		$\epsilon_{600} \times 10^{-3a}$ (M^{-1} cm^{-1})	Redox potential[b] ($E_{1/2}$, mV)	Mediator activity	
				Hydrogen evolution[a]	Hydrogen uptake[b]
				(μmol H_2/min/mg protein)	
R =	R' =				
		Neutral			
	Methylalkyl				
CH_3-	CH_3-	13.0	-443	10,570	1,320
CH_3-	$CH_3(CH_2)_2-$	16.0	—	9,310	2,335
CH_3-	$CH_3(CH_2)_3-$	15.5	—	8,610	840
CH_3-	$CH_3(CH_2)_5-$	16.3	—	8,090	800
	Dialkyl				
$CH_3(CH_2)_2-$	$CH_3(CH_2)_2-$	16.6	-470	9,470	750
$CH_3(CH_2)_4-$	$CH_3(CH_2)_4-$	9.4	—	2,220	500
$CH_3(CH_2)_6-$	$CH_3(CH_2)_6-$	10.4	-385	1,060	1,150
$CH_3(CH_2)_7-$	$CH_3(CH_2)_7-$	8.8	-485	3,750	1,880
	Alkylhydroxy				
$HO(CH_2)_2-$	$HO(CH_2)_2-$	13.0	-399	7,660	1,870
$HO(CH_2)_3-$	$HO(CH_2)_3-$	12.0	—	7,470	1,630
CH_3-	$HO(CH_2)_2-$	12.0	—	9,660	1,790
	Aryl				
CH_3-	$C_6H_5CH_2-$	14.3	-408	9,660	1,930
$C_6H_5CH_2-$	$C_6H_5CH_2-$	11.5	-359	1,280	6,060
		Charged			
	Anionic				
$O_3S(CH_2)_3-$	$O_3S(CH_2)_3-$	14.4	—	540	0
$HOOC(CH_2)_2-$	$HOOC(CH_2)_2-$	12.5	-410	450	0
	Cationic				
$H_2N(CH_2)_2-$	$H_2N(CH_3)_2-$	10.2	-280	2,880	2,830
$H_2N(CH_2)_3-$	$H_2N(CH_2)_3-$	10.5	-390^d	16,960	20,420

[a] Extinction coefficients of the reduced viologen in Tris-HCl (0.1 M, pH 7.4) using a 30-fold excess of sodium dithionite and reducing agent.

[b] One-electron half-wave potentials measured against the normal hydrogen electrode (NHE) taken from Reference 11. Some values given are those initially measured against the saturated calomel electrode and corrected by assuming a potential of $+220$ mv for the saturated calomel electrode SCE vs. NHE.

[c] Hydrogen evolution activity was measured polarographically under standard conditions: sodium acetate buffer (0.1 M, pH 5.5, 37°C), sodium dithionate (10 mM), enzyme (approximately 20 ng), and viologen (3.3 mM). Hydrogen uptake activity was measured spectrophotometrically under standard conditions: Tris-HCl buffer (0.1 M, pH 7.4, 37°C), saturated with hydrogen gas at atmospheric pressure, prereduced enzyme (approx. 20 ng), and viologen (0.33 mM).

[d] This paper measured against the saturated calomel electrode and corrected using a value of $+220$ mv SCE vs. NHE.

Chemical Co. (St. Louis, MO) and sodium dithionite (DTN) was supplied by Anachemia (Montreal, P.Q.). All substituted viologens were obtained from Imperial Chemical Industries Ltd. (Jealot's Hill, U.K.), with the following exceptions: (1) N,N'-dimethyl-4,4'-dipyridyl (methyl viologen) and N,N'-dibenzyl-4,4'-dipyridyl (benzyl viologen) — Sigma Chemical Co., (2) N,N'-diheptyl-5,5'-dipyridyl — Aldrich Chemical Co. (Milwaukee, WI), (3) N,N'-dioctyl-4,4'-dipyridyl (m.p. 300°C), which was prepared by reacting 2 mol of 1-bromoctane (Aldrich) with 4,4'-dipyridyl (Koch-Light Colnbrook, U.K.) in refluxing acetonitrile, and (4) N,N'-diaminopropyl-4,4'-dipyridyl, which was synthesized according to the literature.[10] All viologens were analytical; data corresponded to the given molecular formula within the following limits: C, ±0.2%; H, ±0.2%, N, ±0.3%. Oxidation-reduction potentials were either taken from a compilation[11] or measured by differential pulse polarography (PAR) polarographic analyzer, model 174A) using a standard calomel electrode (SCE) and a dropping mercury electrode, controlled by a PAR drop timermodel 1747. Redox potentials were determined under a nitrogen atmosphere in 0.01 M sodium phosphate buffer at viologen concentrations of approximately $5 \times 10^{-4} M$. Extinction coefficients of the reduced viologens (E_{600}, M^{-1} cm^{-1}) were measured under nitrogen by adding a 30-fold excess of DTN to a known concentration of the viologen in a septum-sealed cuvette and immediately measuring the absorbance at 600 nm (Beckman Spectrophotometer, model DB).

B. MICROORGANISM GROWTH AND ENZYME ISOLATION

D. desulfuricans (ATTC 7757) was grown in medium N and the periplasmic hydrogenase isolated and purified as previously described.[12,13] Crude enzyme isolates were obtained directly after Tris-EDTA extraction of the cells. The purified hydrogenase used in most experiments had specific hydrogen-production activities between 7,500 and 10,000 μmol H$_2$/min/mg protein when a standard assay condition was used. A maximum theoretical value of 45,700 μmol H$_2$/min/mg has been estimated by immunochemical experiments.[14] Enzyme concentration measurements, needed for k$_{cat}$ determinations, were calculated using this value of maximum specific activity and the observed activity of the enzyme as determined under the standard assay condition.

C. ENZYME ACTIVITY DETERMINATION

Hydrogen evolution activity was determined polarographically.[15] Standard reaction mixtures (3 ml), maintained at 37°C, were made up in 0.1 M sodium acetate, pH 5.5, 10 mM DTN, and enzyme (approximately 20 ng). Reactions were initiated by viologen addition to a final concentration of 3.3 mM.

Hydrogen uptake activity was determined spectrophotometrically.[16] Standard reaction mixtures (2 ml), maintained at 37°C, were made up in 0.1 M Tris-HCl, pH 7.5, and 3.3 mM viologen. The reaction mixture was saturated with hydrogen at atmospheric pressure and the reaction was initiated by the

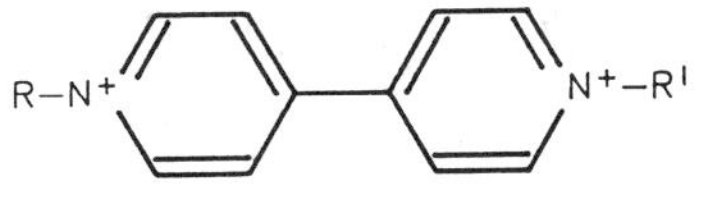

FIGURE 1. Viologen side-chain structures. R and R′ = alkyl and aryl groups, either neutral or charged.

syringe addition of an aliquot of prereduced enzyme. Prereduced enzyme was prepared under nitrogen in 0.1 M Tris-HCl, pH 7.5, containing 0.1% w/v bovine serum albumin and 0.1 mM DTN.

D. KINETIC PARAMETER ESTIMATES

Estimates of median V_{max} (app) and K_m (app) values at the 95% confidence level were obtained from the modified direct linear plot method.[17-19] All initial rate data were also shown to be linear in the Hanes plot (S/V vs. S) format as well as the more usual reciprocal plot (1/S vs. 1/V) format. Estimates of V_{max} and K_m devoid of complicating concentration parameters were made by varying the concentration of both substrates (viologen and hydrogen).[20] Estimates of k_{cat} were made by using the V_{max} values determined previously and estimating the enzyme concentrations from a comparison of the observed activity of the hydrogenase used and the maximum specific activity calculated for the hydrogenase under standardized conditions.[12] The utility of the k_{cat}/K_m (app) parameter for comparing substrates in mono and dual substrate reactions has been previously described.[21-24]

III. RESULTS AND DISCUSSION

A. VIOLOGEN SURVEY

Viologens possessing different side-chain structures (Figure 1) were tested as mediators of both the hydrogen evolution and hydrogen uptake reaction catalyzed by the periplasmic hydrogenase from the sulfate-reducing microorganism *D. desulfuricans*. Both reactions were examined under standard reaction conditions. The hydrogen evolution reaction rates were assayed by a polarographic method in acetate buffer (0.1 M, pH 5.5, 37°C) containing 3.3 mM viologen, while the hydrogen uptake reaction rates were assayed by a spectrophotometric method in Tris-HCl buffer (0.1 M, pH 7.5, 37°C) containing 3.3 mM viologen. Three broad classes of compounds — those with methyl viologen (R = R′ = CH_3, Figure 1). The results are presented in Table 1.

B. NEUTRAL SIDE-CHAIN COMPOUNDS

Viologens with neutral side chains, i.e., methyl-alkyl, dialkyl, alkyl-aryl, and diaryl side chains, were first examined. Interchange of only one methyl group in methyl viologen, to form a series of compounds with methyl-alkyl

groups, resulted in reduced hydrogen evolution and uptake activities except
for the methylpropyl derivative, where hydrogen uptake activity increased.
Among the viologens with dialkyl side chains, it was noted that as the chain
length increased beyond three, a rapid drop in hydrogen evolution occurred.
This could be connected with either their increasing lipophilic nature or their
decreasing accessibility to the viologen binding site. In the hydrogen uptake
reaction, the dipropyl derivative had lowered activity, whereas compounds
with more than three methylene groups were as active as methyl viologen.
Hydroxyl group insertion on the side chains, i.e., to form the alkylhydroxyl
compounds, resulted in a reduction of hydrogen-evolving activity and an
increase in hydrogen uptake activity when compared with their respective
dialkyl counterparts. The one methyl-alkylhydroxyl compound tested had
virtually the same activity in both reactions as its methyl-alkyl counterpart.
The methyl-aryl compound, i.e., methyl-benzyl viologen, had slightly lower
evolution activity, but an increase in the hydrogen uptake reaction when
compared with methyl viologen. The doubly aryl-substituted compound, benzyl viologen, had markedly lower hydrogen evolution activity while showing
an apparent fourfold increase of hydrogen uptake activity. Previous results
have suggested that the lower redox potential of the benzyl viologen (approximately -350 mv vs. -473 mv for methyl viologen) could control the
reaction. The present observations do not seem to support the conclusion that
viologen redox potential exerts total control over both reactions. Indeed, the
redox behavior of the compound seems to be only one of many controls over
both hydrogen evaluation and uptake reactions.

C. ANIONIC SIDE-CHAIN COMPOUNDS

Compounds with anionic side-chain substituents had very low activity in
the hydrogen evolution reaction and none at all in the uptake reaction. The
results with the sulfonatopropyl derivative are in contrast with those obtained
with *Chlamydomonas* hydrogenase and the *Desulfovibrio* hydrogenase used
in a light-driven, chloroplast-coupled hydrogen evolution system.[25-27] The
differences could be attributed to several explanations: (1) different hydrogenases were used, (2) different states of enzyme purity, i.e., presence or
absence of cofactors and mediators, could influence the results, and (3) the
compound enhances coupling between the chloroplasts and the hydrogenase
used.

D. CATIONIC SIDE-CHAIN COMPOUNDS

The most interesting results were obtained with compounds having cationic
side chains. The diaminoethyl derivative was found to behave much like
dibenzyl viologen, since it had lower hydrogen-evolving activity while having
a higher hydrogen uptake activity. Lengthening the alkyl side chains by only
one methylene group to form the diaminopropyl viologen produced a derivative with the highest activity observed regardless of the reaction examined.

This derivative, N,N'-diaminopropyl viologen, was 1.6 times more active than methyl viologen in the hydrogen uptake reaction.

The results of the survey suggest that the viologen active-site area may be restricted in size (compare the methyl-alkyl and dialkyl compounds) and that a charged binding site may be located about 5 Å from the active site (compare the dialkyl and cationic compounds). A fuller description of the groups involved and their implication in the catalytic mechanism will have to await an in-depth investigation of groups involved in the enzyme active site.

Because of its high activity in both the hydrogen production and uptake reactions, the diaminopropyl viologen (Figure 1, R = R′ = $NH_2CH_2CH_2CH_2$–) (DAPV), was subjected to a more thorough investigation. A preliminary account of our results has already been made.[28] By way of summary, it was observed that the k_{cat}/K_m (app) values determined for the diaminopropyl derivative was 1.09×10^{-9} $s^{-1} M^{-1}$ for the hydrogen production reaction and 3.25×10^9 $s^{-1} M^{-1}$ for the hydrogen uptake reaction. Both values indicate that the reaction rate is diffusion controlled, and are substantially higher than the values observed for catalase and carbonic anhydrase (10^7 to 10^8 $s^{-1} M^{-1}$). The values for DAPV are to be compared to the k_{cat}/K_m (app) values obtained for methyl viologen (Figure 1, R = R′ = CH_3^-) (MeV) under the same experimental conditions (i.e., 3.35×10^8 $s^{-1} M^{-1}$, hydrogen production reaction; 2.03×10^8 $s^{-1} M^{-1}$, hydrogen uptake reaction). The results also indicated that a major share of the difference between the two derivatives resides not in differences in k_{cat} values for both reactions, but mainly in differences in the K_m (app) values. DAPV was a much better mediator in all cases, since its K_m (app) was three- to tenfold better than the K_m (app) of MeV. In the absence of hydrogen-gas concentration effects, the K_m (app)s for DAPV and MeV in the hydrogen uptake reaction also differ by a factor of 10 (DAPV = 4.2×10^{-5} M; MeV = 5.7×10^{-4} M).

A K_m (app) difference between the two derivatives also was observed in the hydrogen gas kinetic constants (Table 2) determined in the hydrogen uptake reaction at constant viologen concentration. However, when these hydrogen gas constants were redetermined in the absence of viologen concentration effects, the hydrogen binding constant in the presence of both derivatives seemed to be the same: K_m (app) H_2 (DAPV) = 8.4×10^{-5} M; K_m (app) H_2 (MeV) = 6.7×10^{-5} M (data not shown).

Some of these same observations related to the DAPV have been arrived at by others. Hoogvliet et al.[29,30] recently used cyclic voltammetry and chronoamperometry measurements to assess the calaytic constants for the hydrogenase from *D. vulgaris*. They have suggested the presence of a charged group proximate to the viologen binding site and have also proposed that the electron transfer rate is the rate-determining step in the calaytic mechanism.

Formulation and substantiation of a totally satisfactory mechanistic description of the working of the hydrogenase-mediator system will require, at a minimum, much more precise kinetic data and product inhibition and slope-intercept

TABLE 2
Kinetic Parameters of Hydrogen in the Hydrogenase-
Catalyzed Hydrogen Uptake Reaction
(Constant Viologen Concentration)

	Hydrogen kinetic parameters[a]		
Viologen mediator	K_m (app) (mM)	k_{cat} ($10^5 \cdot s^{-1}$)	k_{cat}/K_m (app) ($s^{-1} M^{-1}$)
Methyl viologen	1.15	2.07	1.38×10^8
Diaminopropyl viologen	0.19	3.5	1.84×10^9

[a]   All values were measured by a spectrophotometric assay under the following conditions: viologen concentration, 3.3 mM; temperature, 37°C; buffer, Tris-HCl pH 8.0, 0.1 M containing bovine-serum albumin (0.5% w/v). Hydrogen concentrations (0.025 to 5.0 mM) were measured polarographically.

analysis, as well as isotope exchange studies. Only when the results of these extensive studies are known will it be possible to propose a reaction sequence which will fully describe the reaction catalyzed by the enzyme hydrogenase. When put in this context, the results presented here are only a small, but, we hope, important step on the way toward this fuller understanding of hydrogenase enzymes.

Finally, on a more practical note, the potent herbicidal action of paraquat, i.e., MeV, has been related to its ability to disrupt the photosynthetic cycle in plants. The present results also suggest that its action could be related, in part, to the discrepancy in the hydrogen evolution and hydrogen uptake activities catalyzed by plant-associated hydrogenases.[31,32] High efficiency in the hydrogen evolution reaction would favor complete exhaustion of redox systems supplying reducing equivalents and poor recovery of the released energy by the back reaction with hydrogen.

ACKNOWLEDGMENTS

The authors wish to express their appreciation to C. J. Dicaire and J. J. Giroux for preparing samples of purified hydrogenase, to K. L. O'Brien, NRC summer student, for preparing large quantities of DAPV, to S. B. Walker, ICI (U.K.) Ltd., for obtaining samples of the many viologens tested, to H. Sequin for elemental analysis, to Dr. R. Renaud for redox potential measurements, and to Dr. A. Storer for advice on the enzyme kinetic treatment.

REFERENCES

1. **Adams, M. W. W., Mortenson, L. E., and Chen, J.-S.,** *Biochim. Biophys. Acta,* 594, 105, 1981.
2. **Voordouw, G.,** in 42nd Symp. of the Society for General Microbiology, 1988, 147.
3. **Hagen, W. R.,** *Photocalalytic Production of Energy-rich Compounds, Eur. 11371,* Commission Eur. Comm., 1988, 241.
4. **Peck, H. D. and Gest, H.,** *J. Bacteriol.,* 71, 70, 1956.
5. **Curtis, W. and Ordal, E. J.,** *J. Bacteriol.,* 68, 351, 1954.
6. **King, N. K. and Winfield, M. E.,** *Biochem. Biophys. Acta,* 18, 431, 1955.
7. **Krasna, A. I. and Rittenberg, D.,** *Proc. Natl. Acad. Sci. U.S.A.,* 42, 180, 1956.
8. **Adams, M. W. W. and Hall, D. O.,** *Biochem. J.,* 183, 11, 1979.
9. **Ikura, I., Aono, S., and Kusunoki, S.,** *Inorg. Chim. Acta,* 71, 77, 1983.
10. **Simon, M. S. and Moore, P. T.,** *J. Polym. Sci. Polym. Chem. Ed.,* 13, 1, 1975.
11. **Bird, C. L. and Kuhn, A. T.,** *Chem. Soc. Rev.,* 10, 49, 1981.
12. **Glick, B. R., Martin, W. G., and Martin, S. M.,** *Can. J. Microbiol.,* 26, 1214, 1980.
13. **Martin, S. M., Glick, B. R., and Martin, W. G.,** *Can. J. Microbiol.,* 26, 1209, 1980.
14. **Ziomek, E., Martin, W. G., and Williams, R. E.,** *Can. J. Microbiol.,* 30, 1197, 1984.
15. **Peterson, R. B. and Burris, R. H.,** *Arch. Microbiol.,* 116, 125, 1978.
16. **Hasche, R. H. and Campbell, L. L.,** *J. Bacteriol.,* 105, 249, 1971.
17. **Eisenthal, R. and Cornish-Bowden, A.,** *Biochem. J.,* 139, 715, 1974.
18. **Cornish-Bowden, A. and Eisenthal, R.,** *Biochem. J.,* 139, 721, 1974.
19. **Cornish-Bowden, A. and Eisenthal, R.,** *Biochim. Biophys. Acta,* 523, 268, 1978.
20. **Segel, I. H.,** *Enzyme Kinetics,* Interscience, New York, 1975.
21. **Northrop, D. B.,** *Anal. Biochem.,* 132, 457, 1983.
22. **Place, A. R. and Powers, D. A.,** *Proc. Natl. Acad. Sci. U.S.A.,* 76, 2354, 1979.
23. **Sudi, J. and Havsteen, B. H.,** *Int. J. Peptide Protein Res.,* 8, 519, 1976.
24. **Fersht, A. R.,** *Proc. R. Soc. London Ser. B,* 187, 397, 1974.
25. **Roessler, P. and Lien, S.,** *Arch. Biochem. Biophys.,* 213, 37, 1982.
26. **Roessler, P. and Lien, S.,** *Arch. Biochem. Biophys.,* 230, 103, 1984.
27. **Cuendet, P. and Grätzel, M.,** *Photochem. Photobiol.,* 36, 203, 1982.
28. **Ziomek, E., Martin, W. G., and Williams, R. E.,** *Ann. N.Y. Acad. Sci.,* 434, 292, 1984.
29. **Hoogvliet, J. C., Lievense, L. C., van Dijk, C., and Veeger, C.,** *Eur. J. Biochem.,* 174, 273, 1988.
30. **Hoogvliet, J. C., Lievense, L. C., van Dijk, and Veeger, C.,** *Eur. J. Biochem.,* 174, 281, 1988.
31. **Akhavein, A. A. and Linscott, D. L.,** *Residue Rev.,* 23, 97, 1968.
32. **Calderbank, A. and Slade, P.,** *Herbicides: Chemistry, Degradation and Mode of Action,* 2nd ed., rev., Kearney, P. C. and Kaufman, D. D., Eds., Marcel Dekker, New York, 1976, 501.

INDEX